DISCRETE SIMULATION AND ANIMATION FOR MINING ENGINEERS

DISCRETE SIMULATION AND ANIMATION FOR MINING ENGINEERS

Editor

Arvind Kumar

scitus
academics

Discrete Simulation and Animation for Mining Engineers
Edited by **Arvind Kumar**

ISBN: 978-1-68117-676-5
Library of Congress Control Number: 2016958831

SCITUS Academics LLC,
616, Corporate Way, Suite 2, 4766,
Valley Cottage, NY 10989

www.scitusacademics.com

Notice

Printed in United States of America on Acid Free Paper

PREFACE

For a number of reasons, the mining industry is moving worldwide to exploit more mineral deposits using underground methods. Due to the lower operating costs associated with economies of scale, large-scale underground block/panel mining methods are becoming very popular. The planning for a caving mine is, however, extremely difficult. Simulation techniques have been used for quite a while by many industries. They have shown that they help the mine planning process, forecast the performance of modeled systems and test alternatives at very low costs. Simulation is a flexible methodology that we can use to evaluate the actions of an existing or possibly proposed business activity, a new product, a production line or possibly an expansion of the plant, etc. By simulating and analyzing the results, we can understand exactly how the current system works and what would happen if we changed it-or perhaps we can estimate how a new system would work. A simulation deals with uncertainty, in the system itself or in the world around it, often but never. Simulation is one of the most popular quantitative methods-because it is so flexible and has many useful results. The use of simulation in project management can illustrate both virtually but realistically the visual display of the project objective and the procedure.

This book investigates the valuable advantages of using a simulation technique for a mining project. Because of its uncertainty, high cost and complexity, mining activities are believed to be particularly vulnerable to more risks than various other business activities. The danger factor is one of the worst things to think about before starting mining projects. Some project risks can be reduced by visualizing and understanding the process before the actual process of initiation. The book illustrated the advantages of computer-aided animation and simulation for a project optimization.

This book offers a comprehensive yet accessible treatment of mine simulation and animation, which is useful in improving the efficiency of industrial mining. This specific guide aims to add to future software development, which provides mining engineers with much more reliable and functional simulation tools. This will require significant safety, environmental and financial progress through improved planning for the development of pre-production in future generations of large, complex underground mines.

TABLE OF CONTENTS

1

DESIGNING A FRAMEWORK TO IMPROVE TIME SERIES DATA OF CONSTRUCTION PROJECTS: APPLICATION OF A SIMULATION MODEL AND SINGULAR SPECTRUM ANALYSIS

Zahra Hojjati Tavassoli [1], Seyed Hossein Iranmanesh [1,*] and Ahmad Tavassoli Hojjati [2]

[1]School of Industrial Engineering, College of Engineering, University of Tehran, Tehran, 11155/4563, Iran

[2]Centre for Transport Strategy, School of Civil Engineering, The University of Queensland, Brisbane, Queensland 4072, Australia

ABSTRACT

During a construction project life cycle, project costs and time estimations contribute greatly to baseline scheduling. Besides, schedule risk analysis and project control are also influenced by the above factors. Although many papers have offered estimation techniques, little attempt has been made to generate project time series data as daily progressive estimations in different project environments that could help researchers in generating general and customized formulae in further studies. This paper, however, is an attempt to introduce a new simulation approach to reflect the data regarding time series progress of the project, considering the specifications and the complexity of the project and the environment where the project is performed. Moreover, this simulator can equip project managers with estimated information, which reassures them of the execution stages of the project although they lack historical data. A case study is presented to show the usefulness of the model and its applicability in practice. In this study, singular spectrum analysis has been employed to analyze the simulated outputs, and the results are separated based on their signal and noise trends. The signal trend is used as a point-of-reference to compare the outputs of a simulation employing S-curve technique results and the formulae corresponding to earned value management, as well as the life of a given project.

Keywords: project management; project progress simulation; singular spectrum analysis; S-curve technique; earned value management

1. INTRODUCTION

Research in the field of Project scheduling mainly consisting of mathematical operations dates back to the late 1950s, and was used to approximately predict the times on when to start and when to finish activities of the project in terms of priority and constraints in resources, while simultaneously pursuing to achieve a particular optimized project objective. Subsequently, a stockpile of research was conducted, accounting for scheduling and planning of projects in different areas (see, for example [1,2,3,4]).

While extensive research has been carried out during the last two decades resulting in models of project scheduling with different features, the bulk of research has mainly focused on fundamental principles of project scheduling. This limitation can possibly be explained by the project schedule failure, due to limited capability, to overcome the uncertainty and different risks which characterize the actual and practical execution of the project.

However, as a significant role in a project life cycle and in its failure or success, project scheduling (in combination with other new techniques) can improve the results [5]. In other words, a construction project schedule has to be considered as a predictive model that can be used with other techniques such as Earned Value (EV) indices to improve resource efficiency calculations, risk analysis, project planning and control and performance measurement.

For more than fifty years, managers have been using Earned Value Management Systems (EVMS) to cope with the complicated task of controlling and adjusting project schedule baseline as the project is being executed, according to total project budget, timed delivery, and account project scope. There is a general consensus over using this well-known management system to use indices like Cost Performance Index (CPI) and Schedule Performance Index (SPI) to account for cost, schedule and technical performance integratively. Besides, this system allows for the calculation of costs, performance indices, schedule variances, forecast project costs and schedule duration. Cioffi proposed previous and new formalism as well as a new notation to be used in earned value analysis [6].

Research in this area has grown slightly, while forecasting techniques and their applications in areas such as price forecasting have grown rapidly. It seems that the lack or inaccessibility of project data is a major barrier to utilizing new forecasting techniques in project management. Applying new forecasting methods requires periodic, progressive time series data, as well as sufficient historical data from similar projects. Hence, in many construction projects, this data is not sufficient to apply forecasting methods and risk analysis. Refer to [5,7,8] for more details.

This paper contributes to serve two objectives. First, it mainly intends to review the literature on project management control research using earned value management and efforts to investigate project time and cost estimates. Although a large number of papers have proposed estimation methods, little attempt has been made to apply these methods to generate project time series data as daily progressive estimations in different project environments. Therefore, a new simulation approach as a novel framework is offered here to reflect the data regarding time-series progress of the project, considering the specifications and the complexity of the project and the environment where the project is performed.

The output can explain the future progress and financial situation of the project at different time scale. This simulator is a powerful tool that can provide daily progressive estimated information which reassures project managers of execution stages of the project although they lack historical data that could help researchers in generating general and customized formulae in further studies.

Secondly, the additional aim of this paper is to use both fictitious and empirical data to obtain and then validate results by using Singular Spectrum Analysis (SSA) and S-Curve techniques. Specifications of this simulator are extracted from the literature or arise from the author's experiences. Thus, researchers can both adjust their ideas to generally investigate project estimations by setting their inputs and study the outputs of the simulator.

This paper has been structured to include the following sections: Section 2 gives brief reviews on project estimates and controls which refer to sources as significant research studies in the related literature; further details are offered on their usage in the experiments related to simulation. In the first subsection, the most important parameters employed in Earned Value Management (EVM) are briefly reviewed, and some resources on the latest literature concerning time and cost performance measurement are displayed. A second subsection deals with the basic concept of SSA approach to analysis and forecast nonparametric time series. The third subsection surveys simulations techniques with reference to previous research studies in the field of project management and its recent advances. Section 3 describes the methodology, the new proposed framework and input and output descriptions and introduces the validation method in three subsections. The simulator has two main inputs described in Section 4. The following section calculates results containing construction project information which are fictitious project data and empirical data. Finally, Section 6 draws conclusions and paves the ground for future research.

2. LITERATURE REVIEW

In this section, we start with a short review of project estimates and controls with references to the earlier most important studies conducted in the literature. The subsections that follow offer further details about the simulation experiments presented in this study, specifically on their usage.

Plans and project schedules are usually developed for each project to make sure that activities are conducted to meet the specifications of the quality expected along with the budget and the allowed time. Many projects lose substantial financial resources every year due to delays in their progress and require additional time and costs for implementation or to reach completion. One of the main concerns of project managers is to fulfill a project within a pre-determined schedule and allocated budget. When differences between planned and actual work performances become considerable, extra management of tasks would be required to bring indices back on course to stay on the planned schedule and budget. During the implementation phase, project progress should be frequently monitored and checked against the planned project schedule so that we can identify and measure any divergences. Exact control of a project depends on suitable and timely access to the project's information, such as the cost and time status of the project.

2.1. EVM

In 1967, the U.S. Federal Government agencies played a significant role in introducing EVM for the first time and using it in programs like large acquisition ones [9]. The U.S. Federal Government was also able to successfully take advantage of EVM in its widespread projects. Project managers could benefit from EVM the most as it keeps them informed of the early indications and procedures of a project's execution and alerts them to eventual corrective action if needed. A large number of research investigations have been devoted to this area, as shown in [5,10], since EVM was introduced. Although EVM was employed to investigate time and cost both, the cost aspect has been the center of attention in most research studies. In a study, EVM was discussed from a price tag perspective [11]. In a later study, it is focused on predicting how long an activity takes by proposing a reliable methodology and using SPI [12] and a formula for estimating project time according to (EDAC) is proposed, i.e., Estimate of Duration at Completion.

During the life cycle of a project, a point of reference is needed for EVM to measure the project's execution periodically. This fixed reference point is supplied by its project's baseline schedule. In order to obtain the time and costs of a project performance, the three key parameters of EVM must be compared to each other and the time and cost performance measures as below:

- Budgeted Cost of Work scheduled (BCWS)
- Actual Costs Work Performed (ACWP)
- Budgeted Cost of Work Performed (BCWP)
- Earned schedule (ES)
- Schedule Variance (SV = BCWP–BCWS)
- Schedule Performance Index (SPI = BCWPBCWS)
- Cost Variance (CV = BCWP–ACWP)
- Cost Performance Index (CPI = BCWPACWP)
- EDAC = PD/SPI

Lipke questioned the classic metrics of SV and SPI previously used since they failed to reliably forecast ending point of time of a project [10]. In this study, however, a time-based measure was designed to cope with false and unreliable behavior, which is dependent on similar principles to those of earned value method, but which renders the money-related metrics into a time aspect. He transformed the time performance metrics SV and SPI to SV(t) and SPI(t) and indicated that they showconsistent and reliable behavior throughout the project life cycle.

EVM was also used for forecasting project duration, comparing the indicators of previous earned value performance, namely SV and SPI, with the recently developed indicators of earned schedule performance SV(t) and SPI(t) and later compared them with the results of time performance measured through the three methods of obtaining time performance in a project [13]. Their results indicated that a general schedule forecasting formula could apply to various project situations.

Later, it was tried to evaluate EV-based methods by reviewing them as they are commonly used to forecast total duration of a project [14]. They delicately controlled the uncertainty level of project to conduct an extensive simulation survey in which they tried to account for the accuracy of forecasts as influenced by the network structure of project, and the reliable and accurate results obtained from the EV-based measures within time perspective. In their study, they investigated the capability of a newly designed method, and the schedule method earned, which hones the relationship between EV metrics and duration forecasts of the project.

A new formula to calculate the (EAC) is presented [15], i.e., Estimate At Completion, of projects to hone EVMS. Their new method contains a formula which is composed of four variables: SPI, Scheduled Percent Complete by Duration (SPCD), Actual Percent Complete by Duration (APCD), and Sum of Durations to Due Time (SDDT).

Studies measuring time performance of the project and/or the concept of earned schedule have been the focus of some publications issued in the popular and academic literature. Empirical data was manipulated by using statistical procedures involved in validation of time performance indices and stability investigations [10].

After all, sensitivity or risk analysis and EVM (whose results were validated using a simulator) were concepts already studied and examined in previous research investigations [5,14,16].

In conclusion, although research in this area has increased only minimally, there is still widespread interest in the need to introduce new indices and formulae into forecasting. Additionally, the use of simulation techniques and simulators has emerged as an effective tool to generate varied experiments.

2.2. SSA

SSA is a component of the more inclusive classification of methods called Principal Component Analysis (PCA), is a powerful technique used to analyze and predict non-parametric time series that include non-stationarity data elements with complex seasonal components and is superior to classical techniques. On the other hand, SSA is a technique used for the analysis of time series that are observed, and it is capable of exposing the main features of the predictability of time series.

The traditional models, namely ARMA, ARIMA, Box–Jenkins and artificial intelligence, such as fuzzy models, etc., demand certain assumptions as restrictive distributional, structural ones and adequate historical data. A new method inspired by genetic colonial theory is introduced to forecast and filter time series [17] This method has many similarity with SSA.

However, SSA does not use any statistical assumptions, namely the series stationarity or residuals normality, and it need not rely on any parametric model to find out the trend or oscillations. Moreover, contrary to many other methods, it is independent of sample size, even small sample sizes.

The theoretical underpinnings and practical background of the SSA technique were introduced by Golyandina et al. in 2001 [18] and further by Hassani [19,20] and Viljoen for common time series [21]. The applications of this technique are many and various, for example, in meteorology, physics, economics, and financial mathematics. In recent years, SSA has been designed and employed to various practical problems, for instance, in industrial production [22], by Afshar et al. [23] to load forecasting in the electricity market as well as other markets, for forecasting CO_2 emissions [24], and for signal extraction in a genetics related application [25]. Moreover, a forecasting method was presented with Artificial Neural Network and SSA [26] and a hybrid model Coupled with SSA was used to forecast rainfall [27]. Finally, SSA is introduced and reviewed the separability techniques and window length [28].

The SSA technique is composed of two steps that are complementary to each other, decomposition and reconstruction.

Figure 1 presents, the standard SSA, which includes two main stages and four subsets. By reconstructing the original series after decomposition, it is possible to remove noise.

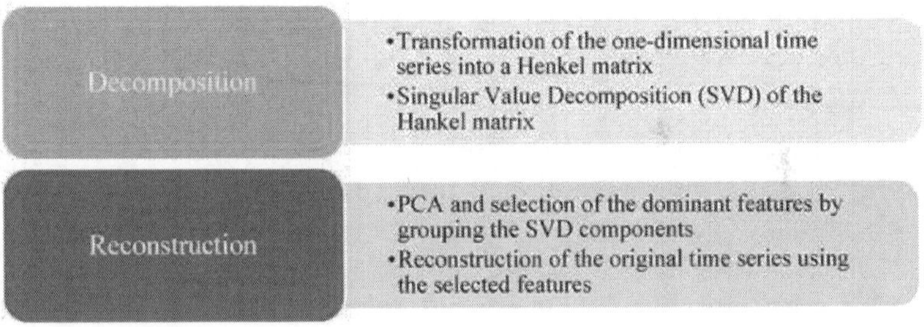

Figure 1. The standard Singular Spectrum Analysis (SSA) steps.

The first stage is the decomposition stage, which has two steps:

Step 1, embedding:

Embedding is similar to mapping and can be regarded as a form of mapping. This form of mapping, i.e., embedding, turns a time series that is one-dimensional, YT=(y1, y2, ..., yT), into series that are multi-dimensional, X1,X2,...,Xk, using vectors, $X_i = (y_i, \ldots, y_{i+L-2}) \in R^L$. .

$$X = \begin{pmatrix} y_0 & \cdots & y_{K-1} \\ \vdots & \ddots & \vdots \\ y_{L-1} & \cdots & y_{T-1} \end{pmatrix}$$

Embedding has certain assumptions: K = T – L + 1; the window length, L, as the single parameter or an integer that should be 2 ≤ L ≤ T. Besides, the L or window length must be big enough. This step results in a matrix called a Hankel matrix, H=(X1,...,XK), which has equal elements on the diagonal, and the linear space that is spanned by columns of H is called the trajectory.

Step 2, SVD:

The second step is called the SVD step, which sums up elementary matrices of rank-one bi-orthogonal to obtain the singular value decomposition of the trajectory, X=E1+...+Ed, where Ei=λi−−√UiVi′ and λi and Ui are eigenvalues and eigenvectors of matrix, $X\acute{X},\ V_i = \acute{X}U_i/\sqrt{\lambda_i}$ and $\lambda_1 \geqslant \lambda_2 \geqslant \ldots \geqslant \lambda_L \geqslant 0.$. The set (λi, Ui, Vi) is referred to as a value of the matrix X called the *i*-th eigentriple, in which s_i represents the singular value of X or its *i*-th value (whose value is equal to the square root of the *i*-th eigenvalue of matrix XX^T); U_i is the *i*-th left singular vector of X (equivalent to the *i*-th eigenvector of XX^T); and V_i is the *i*-th right singular vector of X (equivalent to the *i*-th eigenvector of X^TX). A group of r eigenvectors determines an r-dimensional hyperplane in the L-dimensional space, RL of vectors, Xj.

The second stage is the reconstruction stage, which consists of two steps: grouping and diagonal averaging. The grouping step refers to dividing the elementary matrices, Xi, into some categories and adding the matrices in each group. Imagine that I={i1,...,ip}I={i1,...,ip} is a set of indices i1,...,ip. Therefore, the matrix, XI, related to the set I, is presented as XI=i1+...+ip. The decomposition of the set of indices, J=1,...,d, into the disjoint subsets, I1,...,Im, corresponds to the matrix called X, which is represented by X=XI1+...+XIm. Eigentriple grouping is the process of selecting the sets I1,...,Im.

Diagonal averaging turns every single matrix I into a time series. A time series is an element commonly added to the initial series YT. If zij represents a component of a matrix Z, then the k-th term of the obtained series is calculated by averaging zij over all I, j in such a way that i + j = k + 2. The above procedure is known as diagonal averaging or, to put it differently, the Hankelization of the matrix Z. By applying diagonal averaging or the Hankelization procedure to all matrix elements of XI1,...,XIm, another expansion is achieved: $X = \widetilde{X}_{I_1} + \ldots + \widetilde{X}_{I_m}.$. This expansion is equal to the decomposition of the first original series $y_t = \sum_{k=1}^{m} \widetilde{y}_t^{(k)}$, into a total m series: where $\widetilde{Y_T}^{(k)} = (\widetilde{y_1}^{(k)}, \ldots, \widetilde{y_T}^{(k)})$ k))matches the matrix XIk.

Two parameters are needed for the SSA method: the number of elementary matrices r and the window length *L*. It is necessary that we choose an appropriate window length L since it is essential for improving the accuracy of the SSA. If the points of different vectors, Xi and Xj (i ≠ j) are linearly independent, then the value of L is calculated. The structure of the data and how we perform the analysis determine our choice of these parameters following particular selection rules.

2.3. Simulation Project Techniques

Simulation is also one of the most useful and effective techniques to analyze a complex and dynamic project, which helping the decision maker to recognize the effect of various dependent variables and complex scenarios on the project's behavior. Furthermore, project simulation can be used in different aspects of project management, can help researchers to gain many accurate and realistic outputs according to set inputs and also to investigate different scenarios as summarized below.

As described in Section 2.1, the use of simulation techniques and simulators has emerged as an effective tool to generate varied experiments in this area. A method called Special Purpose Simulation (SPS) is utilized [29] which is a tool to optimize workforce forecast loading and leveling resources. SPS is a simulation model manages to optimize resource supplies and requirements for conducting a petrochemical project, considering standard discipline necessities and involvements. A project model has been designed [30] which simulates the developmental process of the project in a realistic manner and produces information concerning three different monitoring systems. In this paper, factors influencing the total cost and performance of project are also accounted for by changes occurring in the project plan and rates of inflation. In a similar vein, a research framework was developed to enable project managers to approximately calculate completion time of the project and dive deep into the main factors which influence the estimation for complex engineering projects being executed concurrently [31]. Their proposed framework is composed of three main procedures: data gathering, simulation, and data analysis with ANOVA.

A number of studies have been developed to scrutinize various factors affecting project cost estimations. Akintoye focused on understanding different factors influencing project contractors' cost estimating practices [32]. In this study, 24 factors were identified by conducting a comparative study analyzing 84 UK contractors categorized into four classes ranging from very small, small, medium to large firms. Bashir and Thomson presented a thorough quantitative estimation approach which can offer initial along with updated project estimates from manageability or feasibility study until project completion [33]. A dynamic simulation-based crashing methodology was introduced to assess project networks and determine the optimum crashing configuration that reduces the average project cost to a minimum level because of delayed penalties and crashing costs [34]. Their dynamic method allows us to assess the project network to design a crashing strategy at the very onset of the project execution and also during the time that the project is being performed.

As described in Section 2.2, EVM concepts were used as a basis to analyze simulation techniques in a number of research investigations. In a study, the efficiency of controlling a project was analyzed [5]. A Monte-Carlo simulation was used to validate the project data of the study whether fictitious or empirical. Upon noticing warning control parameters, we must take corrective actions to put the project back into the correct path whenever a problem occurs. This set of information offered by the Schedule Risk Analyses (SRA) along with EVM information which is mostly the sensitivity information is obtained during project control. The effects of different activities on risk and the expected completion time of the project were described [8].

In another study, Elshaer used Planned Value Management (PVM), Earned Duration Management (EDM) and Earned Schedule Management (ESM) as the three earned value methods to study the prediction of project duration in EVM [16].

In conclusion, it can be seen that, to a considerable degree, these studies were conducted with the aim of designing a project simulator. In many cases, simulation studies have been used to validate project scheduling solutions or to estimate project specification in some part of the project, for instance project completion cost and time estimations or risk measurements. However, none of the prior studies has considered daily project information as a time series. On this basis, the purpose of this paper is to design a novel project progress simulator that can convert the project into its time series as efficient information, which can help other researchers to generate new formulae.

3. METHODOLOGY

The methodology employed in the computational and conceptual experiment includes a new framework in order to generate project progress time series data, thereby introducing valuable information about projects and preparing an appropriate context for introducing new indices and formulae, varied experiments and further research. In the following subsections, the framework structure, its components and sources and why and how these sources (project progress simulator and SSA) can be used together in project control are explained.

3.1. Project Progress Generator Framework

Computer-assisted simulation is a powerful method to analyze complex and dynamic projects by analyzing large data size, and this precise and flexible technique has been used in much of the research in the field of project management (see, for example, [14,29]). Simulation software can predict the set of expected costs and the duration of the project, depending on input parameters. These inputs control activity duration, cost estimation and the effect of risks when producing cost estimates.

However, as mentioned before, the idea of generating project progress simulators has not been mentioned in the prior studies. Therefore, the aim of this paper is to introduce this simulator and to optimize the resulting time series with SSA. The combination of simulation techniques and SSA analysis with different inputs will help project managers to obtain a realistic estimation of project cost and time in different sectors of the project and will reduce the delay risks. Furthermore, insight is gained into project progress behavior, deviations during project execution, the appropriate use of EAC formulas and identifying areas for additional research.

The important aspect of the SSA is that it has two advantages over other methods, which makes it more beneficial in project control. One feature of the SSA is that it can be applied to series with small sample sizes. The other important aspect is that, unlike other methods, the SSA does not need any statistical assumptions, such as the stationarity of the series or the normality of the residuals. The basic concepts of the project progress generator have been shown by Iranmanesh and Tavassoli [15]. In this paper, the well-developed model with more details is presented. The project progress generator framework has four main steps (as shown in Figure 2). The process starts with the project entry, which can be a given project or selected from a fictitious database and scheduling baseline.

The project progress with the proposed simulator can be considered from two main perspectives:

- Progress Simulation for Time (PST): PST calculates different indices in various time sections and focuses on time parameters to simulate the progress of projects.

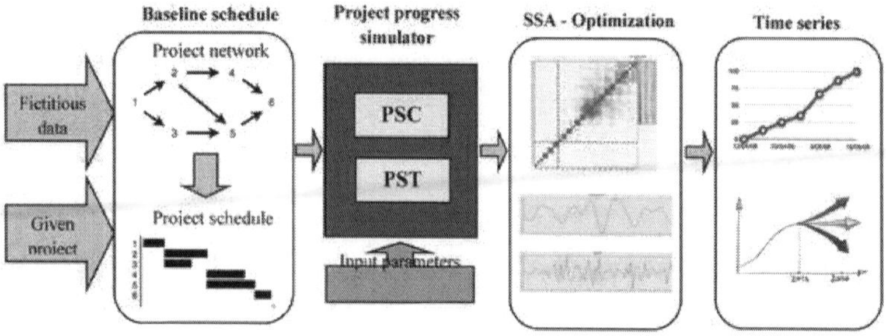

Figure 2. Project progress generator framework and its steps. PSC, Progress Simulation for Time; PST, Progress Simulation for Time.

PST consists of the following steps:

a. Determining the input project and parameters;

b. Constructing project scheduling;

c. Updating the project's progress daily;

d. Computing the project's actual progress daily, based on the set inputs. Suppose $\aleph$ represents the random impact of risk overall, then $\aleph \in (0,1]$. The impact of risk events, X, integrated with the above task progress is shown as follows:

$$(X_i)_j = (X_i)_{j-1} + (P_i)_s * D_i * (\aleph_i)_j$$

where *i* and *j* represent the number of tasks and days, respectively. The random overall impact, $(\aleph_i)_j$, shows the risk of each task in each day from its start to finish. Furthermore, $(P_i)_s$ and $D_i$are daily scheduled progress in a planned day, *s*, and experts' proposed status for the task, *i*.

e.Calculating different measurement indices (i.e., EVM measurements);

f.Continuing until the project is entirely completed.

- Progress Simulation for Cost (PSC): PSC simulates the progress of projects via cost progress during project execution, calculates different indices in different time sections and focuses on cost parameters. PSC is developed mainly based on the PST structure and focuses on the cost parameters. The following is a list of other principles of PSC.

    1. Determining the input costing parameters;

    2. Calculating planned cost progress based on a general specification (i.e., inflation rate) and each task's specifications (i.e., costs of resources);

3. The daily cost of each task is calculated based on resource consumption of the task on that day and other costing parameters;
4. The daily project cost is calculated until the project is finished.

The simulator outputs are time and cost time series that indicate the progressive behavior of the project's cost and time. These outputs are then used as inputs of the SSA method. SSA analyzes these inputs and separates the main trend from noise, then identifies if there is any special trend in the time series, such as seasonal behaviors. The output, or "trend time series", can be used for further estimations if the project is repeated or not completed.

3.2. Input Parameters

The proposed simulator considers several parameters that are identified as affecting the project's progress. Hence, these parameters are presented as samples for the designed model, and many others can be considered in this framework with different types of projects. These input parameters, which should be fed to the simulator by the user, are as follows:

Input project type: As mentioned before, the input project can be a fictitious project or a given one.

Project discipline: In regard to project management, the concept of 'discipline' is defined as an area of work or activity that demands enormous knowledge. It peculiarly corresponds to the area of work or focus where unique and static sets of rules and regulations need to be closely followed in the processes of the conduct and completion of that work or activity.

As Figure 3 presents, the project structure is divided into some separated disciplines, and each discipline comprises certain tasks, each of which utilizes project resources to be accomplished.

Disciplines' work contour: The work contour is a kind of property that demonstrates how every single task is spread over the assignment duration. In this research, it is assumed that all tasks in each discipline have the same work contour property, named back-loaded. The shape of different work contours is pointed out in the different behaviors of tasks and disciplines during the project.

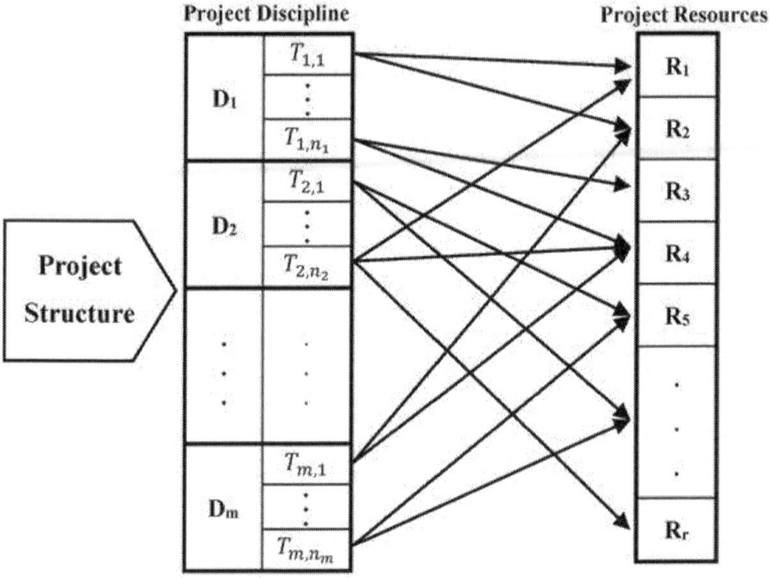

Figure 3. Project disciplines and their relations to project resources.

Expert's comments: To generate actual information, some task specifications are needed. As a case in point, each task's progress status is needed in order to process whether the task progress has overtaken the plan and to determine the degree of variation existing between them. This simulator is a powerful tool for project managers. Therefore, they are the main users of this system, and they should enter their comments about how projects can progress in their environments.

Inflation rate: Inflation rates are assumed to be linear and fixed during the project.

3.3. Output Validation

S-Curve displays accumulated efforts or costs of a project and represents the function of time or cost. In fact, this curve was employed to control and check progress in a set of projects related to each other. Moreover, it is showed that this is possible to characterize the progression of the overall cost of any project [35]. The results of the study inductively concluded that the rising slope in the S-curve, and the point of time when half of the total funds have been consumed, are two numbers that represent this curve.

In the absence of data, we can characterize an equation to illustrate project S-curves with three parameters:

- The total cost of the project (Y_1);
- The total life time of the project (t_1);
- The point of time when the project has spent half of its total funds ($t_{1/2}$).

Estimating the three above parameters, an S-curve equation that can specify in what way labor or funds could be represented during the project at the start of any project can be established for project management purposes.

Define β as the proportion β=t/t1, and term the specific β where Y = $0.5Y_1$ as β1/2 where Y indicates the project cost, given the condition that 0 < β1/2 < 1. If a project whose distribution is front-loaded, where the funds are spent earlier rather than later, β1/2 will be nearer to the beginning of the project. Other projects tend to have this time closer to the middle (β1/2=0.5). For back-loaded projects, in which funds are disbursed nearer to the end of the project, β1/2 would be greater than 0.7.

A gentle rise, a steep slope and a gradual path to the asymptote are the three parts of a typical S-curve. Suppose that the steep slope in the middle portion includes two-thirds of the rise from point Y = 0 to Y = Y∞, where Y∞ is the highest level of the project cost. One-third of the rise remaining takes place over the two gently rising portions. Factor r is defined as a parameter that displays the large size of the rise in the third middle and enables the project manager to freely select an approximation to this central slope, and r0.67 is defined through:

$$r = \frac{2}{3} r_{0.67}$$

Assume that the project manager selected desired β and r0.67, so the value of γ through Equation (1) can be calculated. The project manager can then plot the project's S-curve using Equation (2).

$$\ln(\gamma + 2) = 8\, \beta_{1/2} r_{0.67} \tag{1}$$

Finally, the S-curve equation, in regard to the ratio y = Y/Y_1, is defined as follows:

$$y(\beta) = y_\infty \frac{1 - \exp(-8 r_{0.67} \beta)}{1 + \gamma \exp(-8 r_{0.67} \beta)} \tag{2}$$

where $y_\infty = Y_\infty / Y_1 > 1$. Theoretically, y = 1 shows that the project is complete, and at $\beta = \beta_{1/2}, y = \frac{Y_\infty}{2}$.

If the manager is willing to say that y = 1/2 at β = 1/2, conducting the β′ calculation will obtain a new $\beta_{1/2}$ that will meet the desired situation. Figure 4 represents the flowchart for using this technique to plot a project's S-curve.

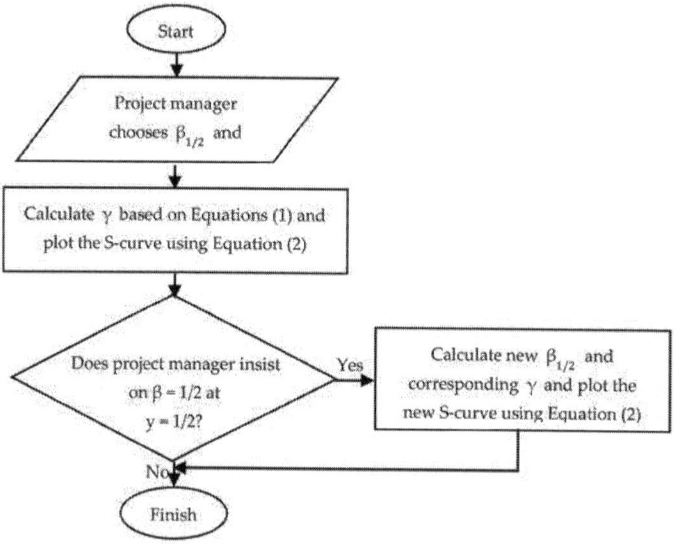

Figure 4. S-curve technique flowchart.

In the next section, the behavior of simulator results is compared to project life cycle trends, using an S-curve trend to verify these results.

4. DATA DESCRIPTION

As mentioned before, the framework has two project data sources: fictitious and given projects. In this subsection, we describe both fictitious and empirical project data and their specifications.

4.1. Fictitious Project Data

As there has been great progress in the models and methods of project scheduling (especially intelligent approaches), the need for data instances to evaluate the solution procedure increases. Therefore, a large number of research studies on project scheduling during the past several years has been attempting to focus on designing and developing benchmark series of project networks to equip users and researchers with successful project planning and evaluating equipment to visualize various projects and to benchmark solution procedures for single and multi-mode project scheduling problems that are resource-constrained.

Generally, if we want to employ intelligent forecasting methods to predict the time and cost of projects, it is essential that we generate basic time series as the scientific representations of various projects. These representations, referred to as standard datasets, are always formed based on the known concept of project networks, explaining the activities, characteristics and precedence constraints of each project using a directed non-cyclic connected graph to prove that the project is manageable.

Comprehensive surveys of these datasets and comparisons between them can be found in [23,25,26,27]. In the current study, we use the ProGen standard dataset. The simulator was used in 480 projects with 92 tasks and 4 resources [36].

As mentioned before, the purpose of this study is to introduce a project progressive simulator framework and generate project time series data. Hence, structural network parameters such as the topological structure, the criticality of a network or resources are not investigated in this study. On this basis, fictitious project data are used just as an input project.

4.2. Empirical Project Data

Javadieh is one of the areas situated in the southwest of Tehran in Iran. An underpass project was conducted in Javadieh. This project includes two bridges, one overpass measuring 90 m in length and 8.20 m in width and one underpass of 80 m in length and 8.20 m in width. Besides, two entry ramps and one exit ramp meet this railway, which cross with the northern city network (Shoosh Avenue) and southern city network (Noori Street).

Engineering, procurement, destruction and construction are the four main disciplines employed in the whole project. The construction discipline consists of some main sub-disciplines, namely excavation, formwork, concreting, waterproofing, embankment, pavement and other minor work activities, like curbs, signs, signaling and lighting.

This project was estimated to be completed within six months with 223 tasks over the four mentioned disciplines. The price agreed for this contract was fixed, approximately 66,869,542,003 Rials (equal to $2.3 million at writing). The project network was thought to be an activity-on-the-node project network whose project data determined the precedence relations between the tasks. The project was scheduled from 24 April 2013 to 20 October 2013; after five months, only 60% of the project had been completed, and the delays made them reschedule the project. Therefore, it was necessary for them to continuously analyze the project condition, measurement factors and daily progress status in order to evaluate the project's efficiency and risk effects. That way, the project can be rescheduled in a realistic way, since this monitoring and controlling equipment enabled the project manager to measure the remaining project cost and time, to provide adequate resources and to take other factors, such as control risks, into account.

5. SIMULATION RESULT

The progress simulator is demonstrated as a software program including PSC and PST concepts by the use of Visual Basic for Application (VBA) in the environment of Microsoft Project 2010 (MSP) and SSA method using CaterPillar SSA 3.40 to analyze the results [37].

In this section, the results of a randomly selected project from the ProGen dataset and a given project are presented. The selected project is just used for presenting the methodology, and this technique can be applied on several projects. Moreover, simulating several runs with different input parameters can help project managers to forecast precisely different situations. The results are then compared to the S-curve technique and SSA, based on the simulator output. The influences of input parameters are then examined in different circumstances. Section 5.1 presents the steps of generating fictitious project progress time series and Section 5.2 demonstrates the simulated results for a given project.

5.1. Fictitious Project Results

Figure 5 shows the results of the simulator on one randomly selected project from the ProGen dataset. In this diagram, the daily progresses of physical and actual percent completion were compared. The S-curve shape can be seen in these two diagrams; however, this simulated project includes a delay of 56 days. This project has 92 tasks and is scheduled for 87 days, and it is simulated by these input parameters:

- The project has four disciplines comprising 20%, 10%, 40% and 30% of all of the tasks, and back-loaded, flat, front-loaded and bell are their work contours, respectively.
- Eighty percent of task delays are in the range of 10% of scheduled progress.
- The project was simulated with a 20% inflation rate for the last six months.

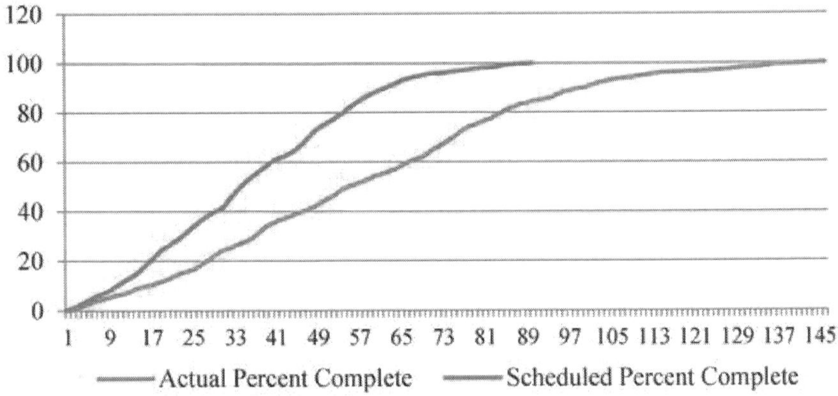

Figure 5. Percent completion of the simulated project.

To validate the simulator, another randomly selected project from the ProGen dataset is used, taking the results and comparing them to the S-curve technique's results. The simulator inputs are set, as all tasks are in one discipline with a normal distribution, a 20% inflation rate for last six months and 80% of task delays in the range of 10% of scheduled progress. Figure 6 shows the actual cost S-curve.

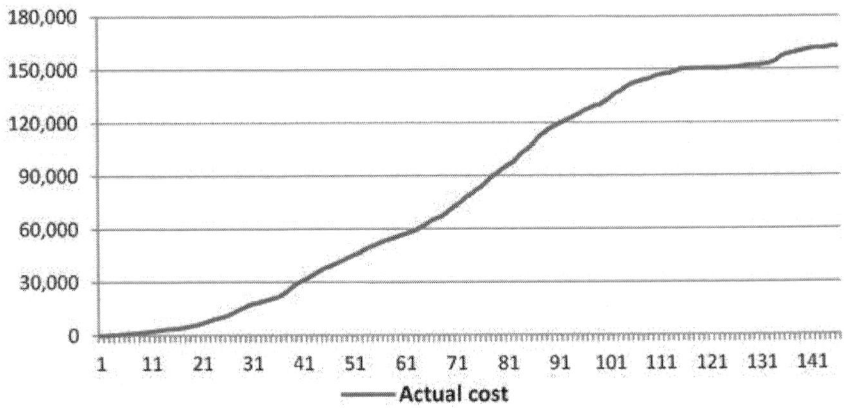

Figure 6. Actual cost S-curve.

This actual time series is now used as an example to illustrate the selection of the SSA parameters and to show the reconstruction of the original series in detail.

- Selection of the window length L:

At the decomposition stage, the window length L is a single parameter that has to be chosen and achieving sufficient separation of the components is crucial in selecting L. As stated above, the window length is the dimension of the Euclidean space that the time series is developed. Moreover, in literature considerable attempt and various methods have been presented for choosing the optimal value of L, for example, $L = T/4$ as a recommended one and L can be larger but not larger than $T/2$ [19]. More recently, the selection of L was optimized based on the minimization of a loss function [38]. Larger value of L leads to a more detailed decomposition and resolving longer period oscillations so L has been set at a larger value ($L = T/2$) in these analyses. In this example $L = 38$.

- Selection of r:

In this method, some auxiliary information and diagrams generated by Caterpillar SSA could be applied to select the parameters L and r. Here, there is a brief explanation of the method which might be beneficial in separating the signal from noise in the time series, which is offered by controlling breaks in the eigenvalue spectra. Furthermore, a pure noise series typically generates a gradually reducing sequence of singular values. It is noteworthy that there are alternate methods for the selection of r, refer to Hassani et al. (2016) for instance [20].

With these accomplished results, the principal components are attained from SVD. Choosing $L = 38$ and performing SVD of the trajectory matrix X, we get 38 eigentriples, organized by the share they contribute to the decomposition. Figure 6 depicts the plan of the logarithms of the 38 singular values. In Figure 7, an enormous decrease in values takes place close to Element 9, which can be taken as the beginning of the noise floor.

Grouping the major elements can be done by another technique in which we analyze the w-correlations of the matrix of the absolute values. Figure 8 represents the w-correlations for the 38 reconstructed elements in a 20-grade grey scale from white to black corresponding to the absolute values of correlations from zero to one. According to this information, the first nine eigentriples (r value) are chosen for the reconstruction of the original series, and the remaining are considered as noise.

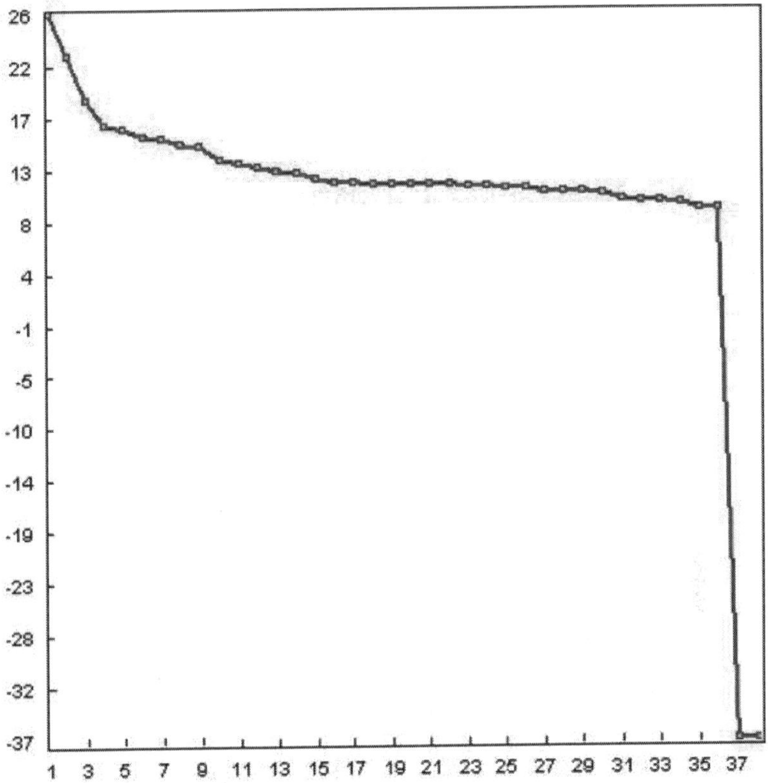

Figure 7. Logarithms of the 38 eigenvalues.

Figure 9 shows the first main elements plotted as time series. Practically, the singular values of the two eigentriples of a harmonic series are usually very near to one another, and this makes the visual identification of the harmonic components easy. A close look at pairwise scatterplots of the singular vectors enables us to visually distinguish those eigentriples that match the harmonic elements of the series, given that these components can be separated from the residual component. Figure 10 presents scatterplots, accompanied by lines joining consecutive points, of the first nine paired major elements.

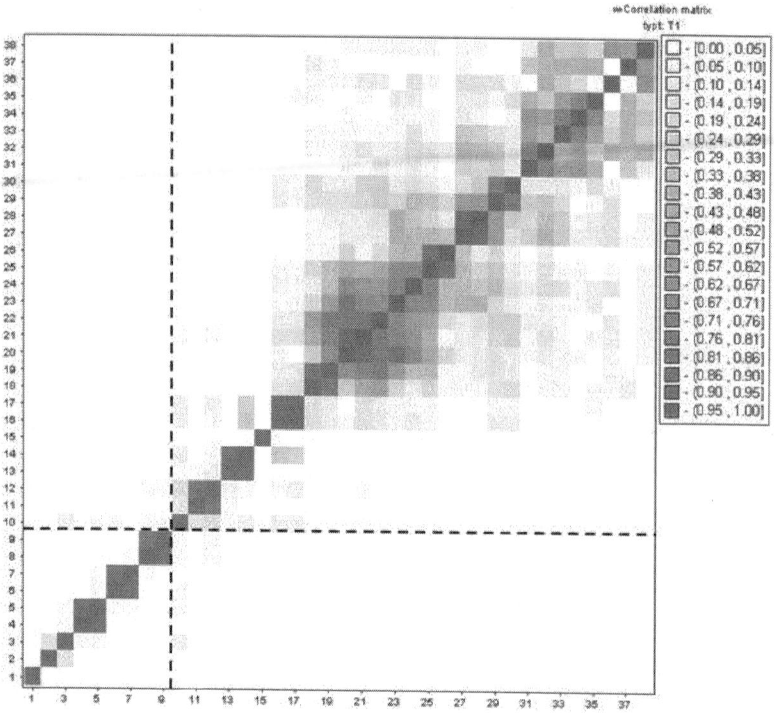

Figure 8. Matrix of w-correlations for the 38 reconstructed components.

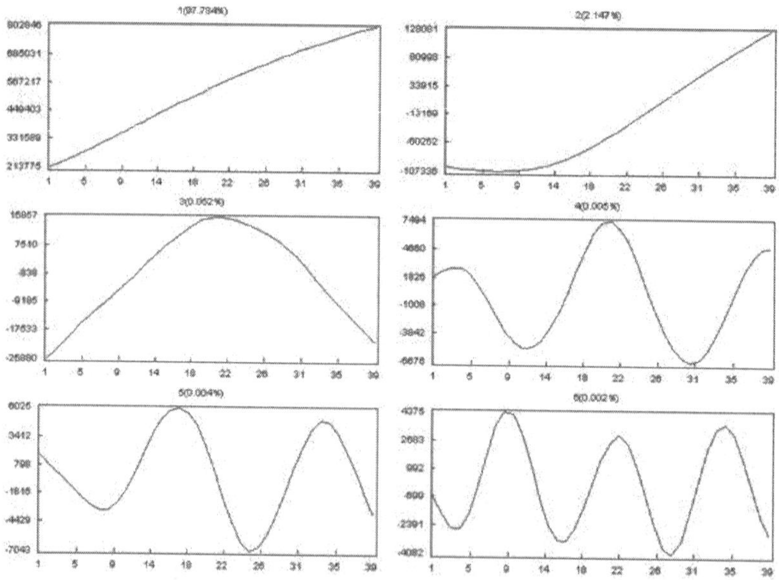

Figure 9. The first six principal components.

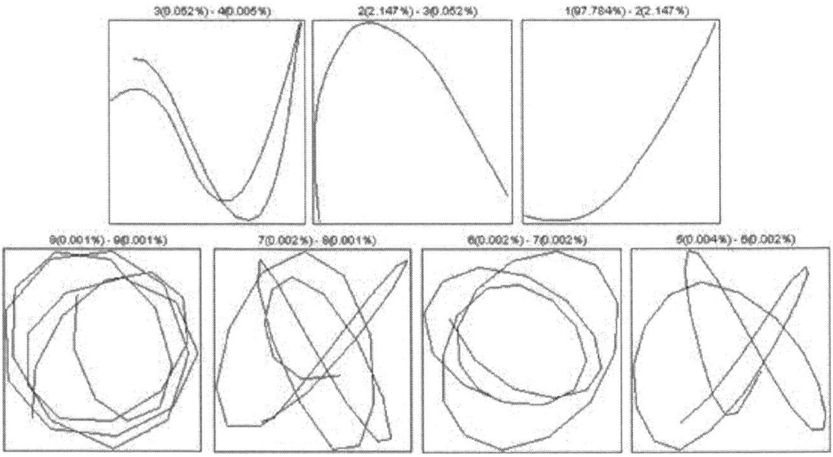

Figure 10. Scatterplots corresponding to the first nine paired principal components.

The trend is a gradual change in the element of a time series that does not include oscillatory components. Thus, to distinguish a trend in the series, one must search for gradually-changing eigenvectors. Figure 11depicts the trend, which is extracted from the eigentriples 1 to 9. This curve explicitly shows the general tendency in the series.

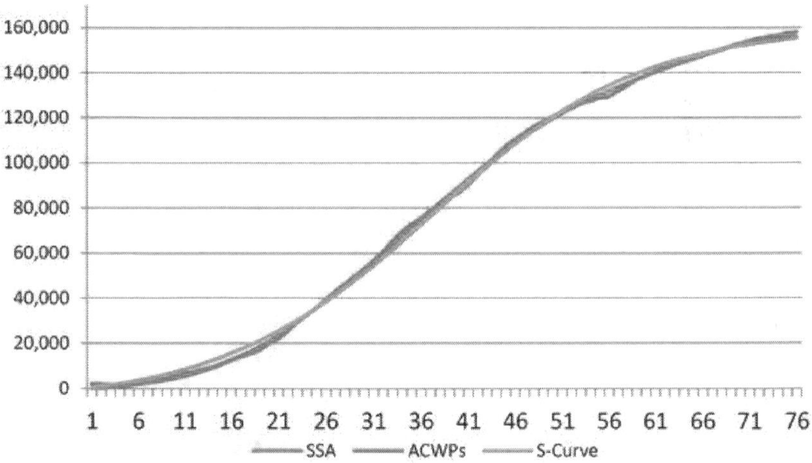

Figure 11. Reconstructed trend from the eigentriples 1 to 9 compared with ACWP and S-Curve.

Figure 12 represents the selected harmonic components (4 to 9) and clearly shows the harmonic behavior of the original series. Therefore, we can categorize the remaining components of the eigentriples (10 to 38) as noise. Figure 13 displays the residuals extracted from these eigentriples.

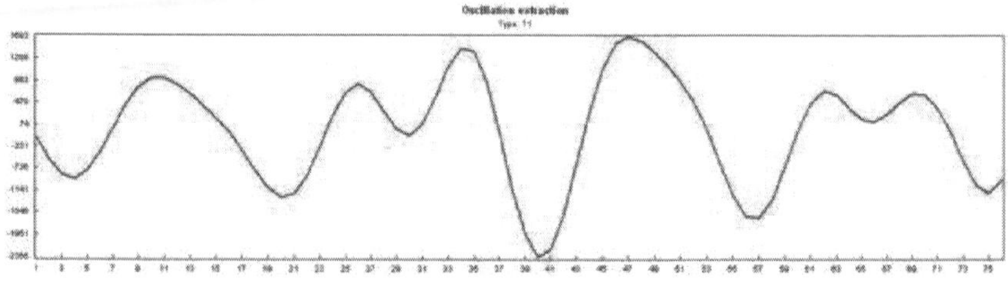

Figure 12. Harmonic from the eigentriples 4 to 9.

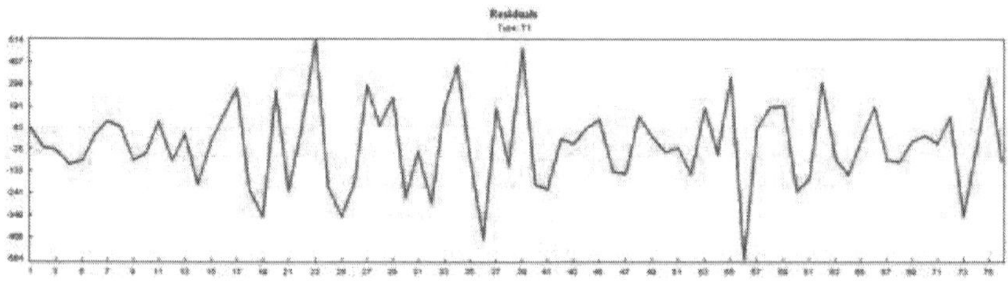

Figure 13. The residuals from the eigentriples 10 to 38.

Finally, as is described in Section 3.3, with chosen values B12/ and r0.67, the parameter γ can be obtained from Equation (1), and Equation (2) can plot the desired evolution S-curve. To fit this curve with an actually simulated curve (the generated time series from the selected fictitious project), we set required parameters as Y_0=100, Y_∞=160,000 and project duration = 77 days. To fulfill the other desired conditions, we set B12/=0.49, γ=29.02 and δ=0.09.

In the next step, the analysis of the S-curve technique time series was implemented similarly to the previously mentioned selected project by SSA method. To show the similarity of these two time series, principal components of two time series, simulator output and S-curve technique output are extracted and compared. The most important components concerning the first three eigentriples, for both the simulator and S-curve technique series, are shown in Figure 14. As shown, these two series have quite similar components. Accordingly, we can conclude that the simulator and S-curve technique time series have the same structure, and this verifies the accuracy of the simulator result.

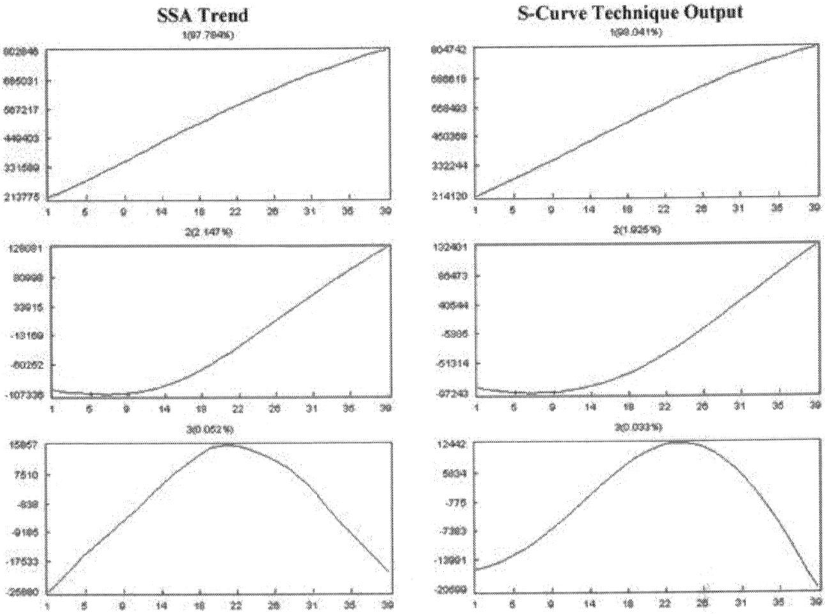

Figure 14. Principal components related to the first three eigentriples.

The deviations between the trend of simulated results (SSA trend) and the S-curve technique with the first generated time series are shown in Figure 15. Clearly, these two deviations have the same behavior over time, with the exception of the start and end points. However, SSA results show a lower change range. Generally, these two deviations are near to one another, which results in the fact that the simulated fictitious results have a high accuracy and that this accuracy can be confirmed by similar tests on empirical data.

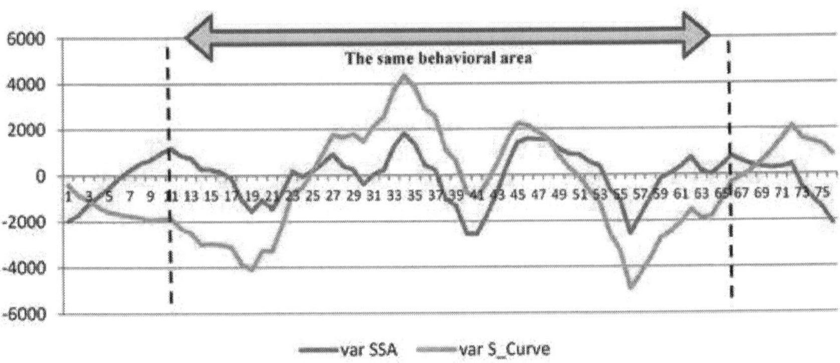

Figure 15. Deviations between the SSA trend and the S-curve method with the generated time series.

To validate deviations between SSA and the S-curve method, two measures, RMSE and R, are employed that assess the quality of the generated time series. These two measures, R and RMSE, are used to determine the extent to which the model fits the data and are calculated as below:

$$R = \frac{\sum (y_i - \overline{y}_i)(\hat{y}_i - \overline{\hat{y}}_i)}{\sqrt{\sum (y_i - \overline{y}_i)^2 \sum (\hat{y}_i - \overline{\hat{y}}_i)^2}} \qquad RMSE = \sqrt{\frac{1}{N}\sum (y_i - \hat{y}_i)^2} \tag{3}$$

Table 1 presents the results of the two measures for SSA trend and the S-curve method with the generated time series. These two measures prove that the generated time series fit the actual trend better than the standard S-curve. Both RMSE values are less than 5% of the maximum member of each series that prove the validity of estimated time series. RMSE of the SSA trend is less than half of the S-curve one. Although the R measure is perfect in both, SSA is judged slightly better.

Table 1. Deviations between SSA trend and the S-curve method with the generated time series.

	MAX	RMSE	R
ACWP	156,015.81		
SSA	158,153.98	1050.64	0.92
S-Curve	155,155.24	2166.50	0.93

5.1.1. Effect of Input Parameters on the Simulator Results

The sensitivity analysis of some important input parameters is presented in this section. Resource constraints, work contours and changes in the progress status are considered as changing factors, and their effects on simulator outputs are studied.

- Effect of Resource Constrains:

Taking into consideration resource constraints, both the project time and cost will increase. Figure 16shows the resulting S-curve of the project for both with and without resource leveling.

Resource leveling lengthens the project completion time 40 days and increase the project cost by about $9500. Moreover, simulator output in the case of resource leveling fluctuates considerably. These two time series have been analyzed by SSA. The window length is set to half of the project duration (L = 53). Figure 17shows the principal components related to the first three eigentriples of the resulting time series.

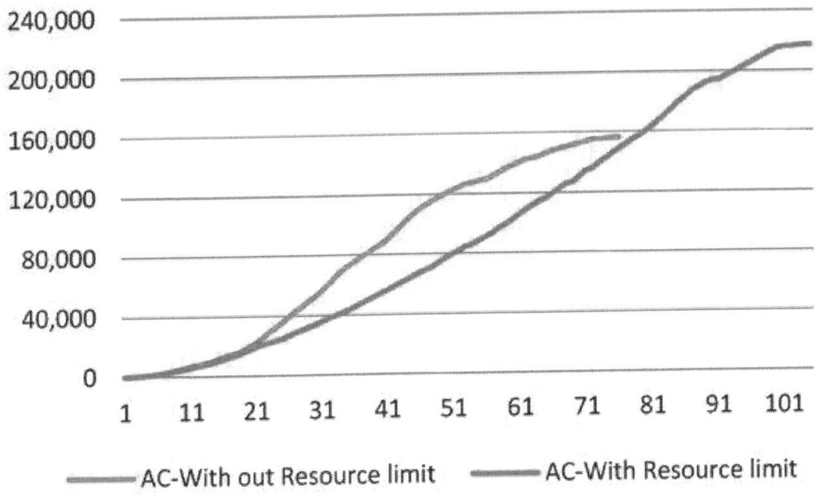

Figure 16. Simulator output with and without resource leveling. AC, Actual Cost.

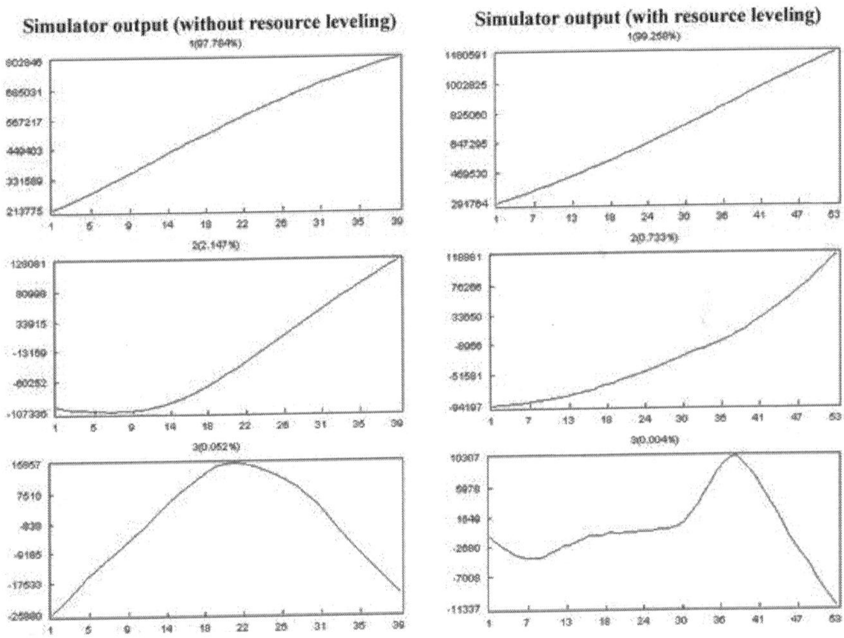

Figure 17. Principal components related to the first three eigentriples.

The two first principal components have approximately the same behavior, and in fact, both series have a similar trend. However, the third principle components of new time series have a different structure due to the high fluctuation of the series. In the case of resource leveling, based on w-correlations analysis, no harmonic components are detected in the resulting time series and residuals, which are obtained by the eigentriples 4 to 53 fluctuating more.

- Effect of discipline and work contour:

In this section, the simulator is run in two different settings of work contour: front loaded and back loaded and with the assumption of resource leveling. Principal components related to the first three eigentriples of the resulting time series are presented in Figure 18.

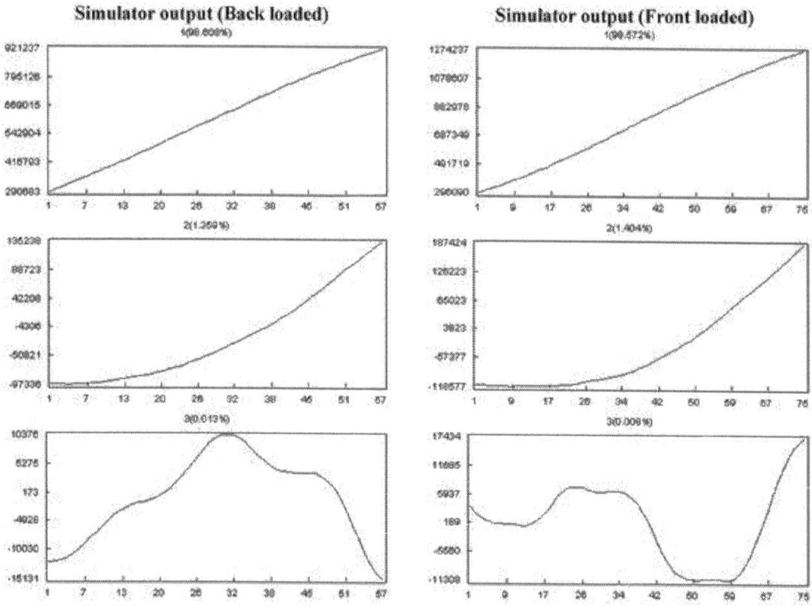

Figure 18. Principal components related to the first three eigentriples of the time series in two different settings: front-loaded and back-loaded work contour.

In this case, the two first components of the resulted series are similar, but the third component has a different shape. Besides, the third component has more fluctuation in the front-loaded work contour setting than the back-loaded one.

Although the residuals of these time series have various behaviors and structures, these series have a similar harmonic component, as well.

- Effect of the changes in the progress status:

To evaluate the impacts of activities' progress status on actual progress and the simulator output, the simulator is run in two various statuses:

- The actual progress of tasks fluctuates ±0.2 of their plan progress.
- The actual progress of tasks fluctuates ±0.3 of their plan progress.

Moreover, it is assumed that the project has a flat work contour and resource leveling. When the interval range increases, project completion time and cost raise, while project progress fluctuates more.

Figure 19 presents the first three eigentriple components of the resulting time series. The increase of this range results in a large fluctuation in the time series. This is apparent in the second and third principle components. However, the first principle component of the two series has approximately the same structure, and therefore, the two series have generally similar trends.

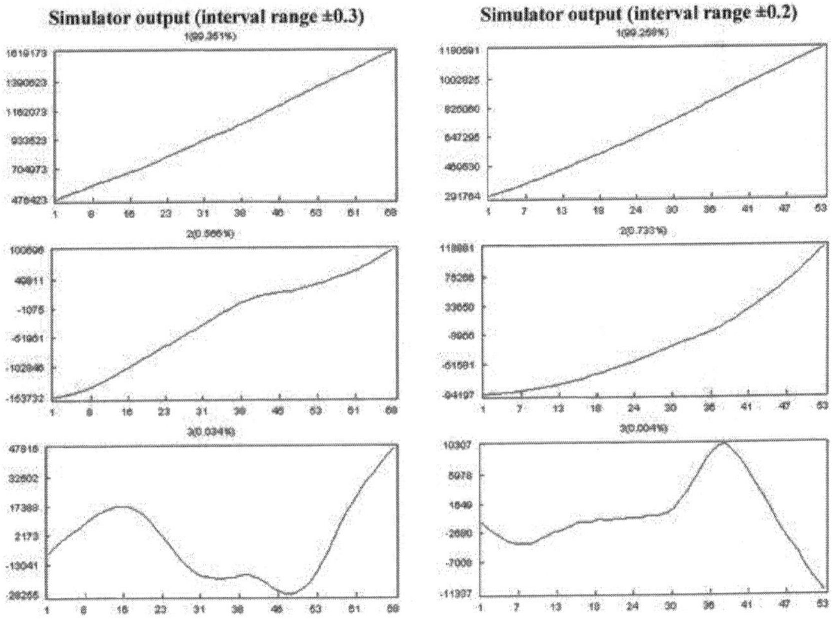

Figure 19. Principal components related to the first three eigentriples of the time series for two different interval ranges.

5.2. Case Study Results

Figure 20 simulates an experiment on a construction project whose project duration was six months, and its Budget at Completion (BAC) was = 66,869,542,003 Rials (around $2.3 million at writing). About 34% of the project was not completed due to delays for about five months since the project started, and the project was executed under a situation with high risks; the inflation rate was estimated to be 30% for six months. As already stated, the project was supposed to follow the four main disciplines and seven sub-disciplines corresponding to the fourth main discipline. Here, we suppose that all tasks have a uniform (flat) distribution.

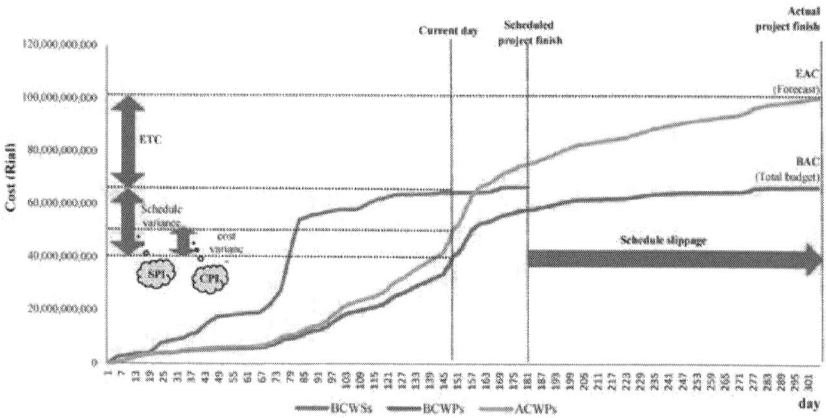

Figure 20. Simulation results for the given project. EAC, Estimate At Completion; BAC, Budget At Completion; SPI, Schedule Performance Index; CPI, Cost Performance Index.

The baseline schedule is used when the project was planned at the very beginning, which is shown in the Gantt chart formed. In this paper, the simulation is conducted by the use of the project progress simulator described earlier. After all, other input parameters have been treated based on present actual information. After 150 days of the project have passed, the simulation begins to match the actual curve with actual project progress; thus, the completion trends can be forecasted and measured.

As seen in Figure 20, the project is scheduled to be completed in 306 days (about 10 months) with a cost overrun. As a result, the budget must rise from 66,869,542 to 100,750,109,952 Rials, i.e., by about 35%. Besides, Figure 20 displays the progress information of another project. Therefore, EVM calculations are now able to be measured daily.

The progress simulator equips us with forecast and control methods that lead to reliable results, and which make daily and completion date forecasting simple to a large extent. If these initial warning signals are appropriately analyzed properly in comparison with the project schedule, they pave the way for us to take corrective actions on those activities that are not on the right track (particularly, those tasks that are on the critical path).

As mentioned before, the generated time series can be used to extend formulae. Table 2 shows the time series analysis with EVM formulae in different time scales of the forecasted time series.

Table 2. Time series analysis with EVM formulae. EDAC, Estimate of Duration at Completion.

Time of Forecast	% Schedule Completed	% Work Completed	EDAC	ES	CV%
1/00	0/32	0/06	225/90	0/03	0
32/00	11/36	17/91	383.08	0/76	0/34
63/00	35/78	31/84	584.70	1/88	0/92
94/00	79/50	45/16	762.57	0/45	3/17
125/00	92/93	53/02	448.16	0/53	8/04
156/00	95/39	62/80	241.40	0/70	18/94
187/00	100/00	73/21			27/44

Figure 21 illustrates one sample of the usage of the simulator results. In this figure, the dynamic action threshold on the Schedule Performance Index (SPI(t)) is depicted.

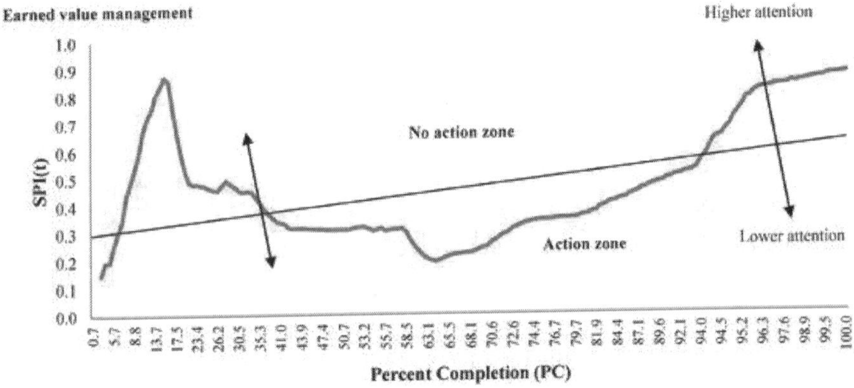

Figure 21. Corrective action during project control.

6. CONCLUSIONS

Construction project estimation and control as important parts of project management and performance measurement are not new to and have long been discussed in the academic research literature. Moreover, project time and cost estimation, especially at the beginning of projects, is one of the most significant processes of project management and, when performed well, results in project risk reduction.

Plenty of research has been completed to estimate project time and cost, most approaches of which are based on EVM techniques and indices, such as SPI, CPI and their combinations. However, applying new forecasting methods requires periodic and progressive time series data, as well as the availability of historical data from similar projects. In this regard, for many projects, the available data are not sufficient to apply forecasting methods.

In this paper, a new simulation approach as a novel framework is offered here to serve as an attempt to introduce a new simulation approach to reflect the data regarding the time series progress of the project, considering the specifications and the complexity of the project and the environment where the project is performed. This simulator is a powerful tool that can provide daily progressive estimated information for project managers to assist their monitoring of project performance, though they the lack historical data that could help researchers in generating general and customized formulae in further research.

This simulator generates project time series with both the PSC and PST concepts and as a part of the presented framework, its outputs are analyzed by SSA. The application of the SSA technique in the area of project management and project control is presented for the first time by this research. The results of the analysis show the superiority of this technique in this field, the specifications of project time and cost curves and the effects on different situations. Furthermore, the results of one random project from the ProGen dataset (as a fictitious input) demonstrate the similarity of the first two components of each curve and other changes according to project specification change.

The simulator's validity is tested through comparison of its output with the results of the S-curve technique, using one of the most powerful tools of analyzing time series, SSA. This comparison demonstrates similar behavior within these tools.

Finally, the result of the simulator is shown with empirical data input. A given project is used as a simulator input, and its time series is then depicted in a similar process. The presented results show that the simulator can be used to forecast the actual costs and the duration of a project, in order to review the planning not only at the beginning, but also while the project is being executed and can help project managers to perform corrective actions. In addition, the project behavior can be investigated during different project periods.

Given the importance of understanding and improving project progress information and data estimations, future lines of research should be set up. The authors intend to focus on two research lines. First, the proposed simulators can be developed for an SPS through adding specific project characteristics, such as resource types. This method allows project managers to generate a more specific time series, which shows project features more precisely. Secondly, the proposed simulator uses other network generators' outputs to simulate a project. Since all datasets have been generated in the resource constraint environments and other assumptions limiting the structure of their networks, the general results cannot be evaluated. Therefore, adding a network simulator, which allows project managers to study network parameter effects on project time and cost, could also be a field of future study.

ACKNOWLEDGMENTS

The authors would like to extend their deepest gratitude to "Gozarrah Consulting Engineers Company" for their support and providing invaluable information and empirical project data in this research. We also acknowledge experts in "Research Institute for Energy Management and Planning" (RIEMP) affiliated to University of Tehran for their sincere and whole-hearted support in all steps of this study.

AUTHOR CONTRIBUTIONS

This paper has been extracted from the thesis of Zahra Tavassoli Hojjati under the supervision of Seyed Hossein Iranmanesh at University of Tehran and invaluable comments of Ahmad Tavassoli Hojjati as co-supervisor at University of Queensland. All stages of this research have been thoroughly checked by the supervisor and co-supervisor of the project.

CONFLICTS OF INTEREST

The authors declare no conflict of interest.

REFERENCES

1. Kolisch, R.; Sprecher, A. PSPLIB—A project scheduling problem library. Eur. J. Oper. Res. 1996, 96, 205–216.
2. Kolisch, R.; Padman, R. An Integrated Survey of Project Management Scheduling; Technical Report; Heinz School of Public Policy and Management, Carnegie Mellon University: Pittsburgh, PA, USA, 1997.

3. Hartmann, S.; Briskorn, D. A survey of variants and extensions of the resource-constrained project scheduling problem. Eur. J. Oper. Res. 2010, 207, 1–15.

4. Kolisch, R.; Hartmann, S. Experimental investigation of heuristics for resource-constrained project scheduling: An update. Eur. J. Oper. Res. 2006, 174, 23–37.

5. Vanhoucke, M. Measuring the efficiency of project control using fictitious and empirical project data.Int. J. Proj. Manag. 2012, 30, 252–263.

6. Cioffi, D. Designing project management: A scientific notation and an improved formalism for earned value calculations. Int. J. Proj. Manag. 2006, 24, 136–144.

7. Iranmanesh, S.H.; Kimiagari, S.; Mojir, N. A new formula to estimate at completion of a project's time to improve earned value management system. In Proceedings of the IEEE International Conference on Industrial Engineering and Engineering Management (IEEM), Singapore, 2–4 December 2007; pp. 1014–1020.

8. Madadi, M.; Iranmanesh, H. A management-oriented approach to reduce project duration and its risk (variability). Eur. J. Oper. Res. 2012, 215, 751–761.

9. Fleming, Q.W.; Koppleman, J.M. Earned Value Project Management; Project Management Institute: Newtown Square, PA, USA, 1996.

10. Lipke, W.; Zwikael, O.; Henderson, K.; Anbari, F. Prediction of project outcome: the application of statistical methods to earned value management and earned schedule performance indexes. Int. J. Proj. Manag. 2009, 27, 400–407.

11. Fleming, Q.; Koppelman, J. What's your project's real price tag? Harv. Bus. Rev. 2003, 81, 20–21.

12. Jacob, D.S.; Kane, M. Forecasting schedule completion using earned value metrics. Meas. News 2004, 3, 11–17.

13. Vandevoorde, S.; Vanhoucke, M. A comparison of different project duration forecasting methods using earned value metrics. Int. J. Proj. Manag. 2006, 24, 289–302.

14. Vanhoucke, M.; Vandevoorde, S. A simulation and evaluation of earned value metrics to forecast the project duration. J. Oper. Res. Soc. 2007, 58, 1361–1374.

15. Iranmanesh, S.H.; Tavassoli, H.Z. Intelligent systems in project performance measurement and evaluation. In Engineering

Management, intelligent systems references library 87; Springer International Publishing: Basel, Switzerland, 2015.

16. Elshaer, R. Impact of sensitivity information on the prediction of project's duration using earned schedule method. Int. J. Proj. Manag. 2013, 31, 579–588.

17. Hassani, H.; Ghodsi, Z.; Silva, E.S.; Heravi, S. From nature to maths: Improving forecasting performance in subspace-based methods using genetics Colonial Theory. Dig. Signal Process. 2016, 51, 101–109.

18. Golyandina, N.; Nekrutkin, V.; Zhigljavsky, A. Analysis of Time Series Structure: SSA and Related Techniques; Chapman & Hall/CRC: Boca Raton, FL, USA, 2001.

19. Hassani, H. Singular Spectrum Analysis: Methodology and comparison. J. Data Sci. 2007, 5, 239–257.

20. Hassani, H.; Zokaei, M.; von Rosen, D.; Amiri, S.; Ghodsi, M. Does noise reduction matter for curve fitting in growth curve models? Comput. Methods Progr. Biomed. 2009, 96, 173–181. [PubMed]

21. Viljoen, H.; Nel, D.G. Common singular spectrum analysis of several time series. J. Stat. Plan. Inference2010, 140, 260–267.

22. Hassani, H.; Heravi, S.; Zhigljavsky, A. Forecasting European industrial production with singular spectrum analysis. Int. J. Forecast. 2009, 25, 103–118.

23. Afshar, K.; Bigdeli, N. Data analysis and short-term load forecasting in Iran's electricity market using singular spectral analysis (SSA). Energy 2011, 36, 2620–2627.

24. Silva, E.S. A combination forecast for energy-related CO2 emissions in the United States. Int. J. Energy Stat. 2013, 1, 269–279.

25. Ghodsi, Z.; Silva, E.S.; Hassani, H. Bicoid signal extraction with a selection of parametric and nonparametric signal processing techniques. Genom. Proteom. Bioinform. 2015, 13, 183–191. [PubMed]

26. Wu, C.L.; Chau, K.W. Rainfall-Runoff Modeling Using Artificial Neural Network Coupled with Singular Spectrum Analysis. J. Hydrol. 2001, 399, 394–409.

27. Chau, K.W.; Wu, C.L. A Hybrid Model Coupled with Singular Spectrum Analysis for Daily Rainfall Prediction. J. Hydroinform. 2010, 12, 458–473.

28. Hassani, H.; Mahmoudvand, R.; Zokaei, M. Separability and window length in singular spectrum analysis. C. R. Acad. Sci. Paris Ser. I. 2011, 349, 987–990.

29. Hanna, M.; Ruwanpura, J.Y. Simulation tool for manpower forecast loading and resource leveling. In Proceedings of the 2007 Winter Simulation Conference, Washington, DC, USA, 9–12 December 2007; pp. 2099–2103.

30. Al-Jibouri, S.H. Monitoring systems and their effectiveness for project cost control in construction. Int. J. Proj. Manag. 2003, 21, 145–154.

31. Huang, E.; Chen, S.J. Estimation of project completion time and factors analysis for concurrent engineering project management: A simulation approach. Concurr. Eng. Res. Appl. 2006, 14, 329–341.

32. Akintoye, A. Analysis of factors influencing project cost estimating practice. Constr. Manag. Econ. 2000,18, 77–89.

33. Bashir, H.A.; Thomson, V. Project estimation from feasibility study until completion: A quantitative methodology. Concurr. Eng. Res. Appl. 2001, 9, 257–269.

34. Kuhl, M.; Radhames, A.; Pena, T. A dynamic crashing method for project management using simulation-based optimization. In Proceedings of the 40th Conference on Winter Simulation, Austin, TX, USA, 7–10 December 2008; pp. 2370–2376.

35. Cioffi, D.F. A tool for managing projects: An analytic parameterization of the S-Curve. Int. J. Proj. Manag. 2005, 23, 215–222.

36. The ProGen standard dataset. Available online: http://129.187.106.231/psplib/main.html (accessed on 13 July 2016).

37. Caterpillar-SSA Software. Available online: http://www.gistatgroup.com/cat/index.html (accessed on 13 July 2016).

38. Hassani, H.; Webster, A.; Silva, E.S.; Heravi, S. Forecasting U.S. tourist arrivals using optimal Singular Spectrum Analysis. Tour. Manag. 2015, 46, 322–335.

2

A STOCHASTIC SIMULATION FRAMEWORK FOR TRUCK AND SHOVEL SELECTION AND SIZING IN OPEN PIT MINES

S.R. Dindarloo[I]; M. Osanloo[II]; S. Frimpong[I]

[I]Department of Mining and Nuclear Engineering, Missouri University of Science & Technology, Bella, MO

[II]Department of Mining and Metallurgical Engineering, Amirkabir University of Technology, Tehran, Iran

SYNOPSIS

Material handling in open pit mining accounts for about 50% of production costs. The selection and deployment of efficient, safe, and economic loading and haulage systems is thus critical to the production process. The problems of truck and shovel selection and sizing include determination of the optimal number and capacities of haulage and loading units, as well as their allocation and operational strategies. Critical survey and analysis of the literature has shown that deterministic, stochastic, and experimental approaches to these problems result in considerably different outputs. This paper presents a comprehensive simulation framework for the problem of truck and shovel selection and sizing based on the random processes underlying the network-continuous-discrete event nature of the mining operation. The framework builds on previous research in this field and attempts to address limitations of available methodologies in the form of a comprehensive algorithm. To test the validity of the framework a large open pit mine was evaluated. The stochastic processes governing the uncertainties underlying the material loading and haulage input variables were defined and built into the stochastic model. Discrete event simulation was used to simulate the stochastic model. The proposed model resulted in several modifications to the case study.

Keywords: truck and shovel operation, stochastic simulation framework, equipment selection and sizing.

INTRODUCTION

There are three different mechanisms of discrete, continuous, and hybrid material transportation in open mines, including (but not limited to) (i) truck and shovel system (discrete), (ii) slurry piping (continuous), and (iii) in-pit crusher and conveyor belts (hybrid). Although each mechanism has its own advantages, the truck and shovel system is the dominant method of material loading and handling in open pit mines, owing to its high production rate, excellent flexibility, relatively low operating and capital costs, and good maintainability. General objectives of optimal equipment selection include (i) meeting the long- and short-term requirements of production rates, (ii) human and equipment safety, (iii) environmental protection, and (iv) economic operations (Figure 1).

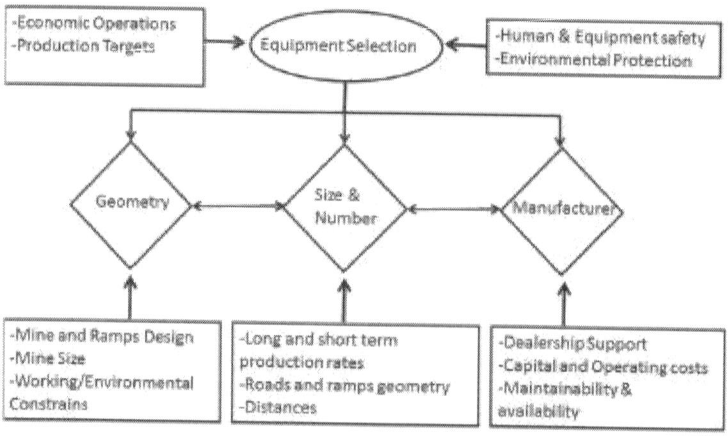

Figure 1—Essentials of equipment selection

The two most important decision factors regarding selection of a truck and shovel system are the equipment geometry and size.

Geometry (equipment width, weight, turning radius, swing angle, *etc.)* is controlled mostly by mine design and layout, as well as the operational constraints. After selection of a favorable geometry based on the constraints, the next step is to select the equipment manufacturer. Once the geometry and manufacturer are selected, the next step is to decide on the models (bucket size) and required numbers of each unit.

Because of the undeniable effect of a proper truck-shovel system selection and sizing plan on open pit mines economics, many researchers have tried to study this issue using different techniques such as linear programming (Edwards, Malekzadeh, and Yisa, 2001), analytical hierarchy process (Ayağ, 2007), nonlinear programming (Søgaard and Sørensen, 2004), genetic algorithms (Aghajani, Osanloo, and Akbarpour, 2007; Marzouk and Moselhi, 2003), mixed integer programming (Camarena, Gracia, and Cabrera Sixto, 2004), machine repair modelling (Krause and Musingwini, 2007), queuing theory (Komljenovic, Paraszczak, and Fytas, 2004), and conventional spreadsheet calculations based on experience, engineering judgment, and manufacturers' catalogues (Burt *et al.,* 2005). Due to the large number of parameters that affect the system performance and the stochastic nature of the input variables, developing a deterministic mathematical optimization solution for the problem is extremely difficult, if not impossible (Haldar and Mahaderan, 2000).

A truck and shovel operation is a set of discrete-event activities, *i.e.* loading, hauling, dumping, and returning, which all occur in a stochastic manner (Figure 2). Discrete-event system simulation (DES) is a modelling method for such time-discrete and probabilistic phenomena (Schriber, 1992). Other abovementioned techniques have different limitations in addressing this problem comprehensively and accurately (Schriber, 1991). These limitations include deterministic pre-assumptions and/or not considering the real-world system specifications through derivation and application of relevant time-frequency distributions for the different operations involved. Most of these techniques therefore do not lead to robust models (Burt *et al.,* 2005). However, DES has been employed by different researchers in mining engineering through available software and languages such as GPSS, SIMAN-ARENA, and SLAM (Baffi and Ataeepur, 1996; Runciman, Vagenas, and Newson, 1996; Awuah-Offei, Temeng, and Al-Hassan, 2003; Ross *et al.,* 2010; Sturgul, and Thurgood, 1993). Most of the studies to date have endeavoured to evaluate some what-if scenarios in order to understand the possible effects of changing different input variables on the overall economics of current operating mines. For instance, (Stout *et al.,* 2013) used Arena to simulate a truck and shovel operation. Very good background reviews of the application of this technique in the mining industry can be found in Sturgul, (1995, 1999) and Hollocks, (2006).

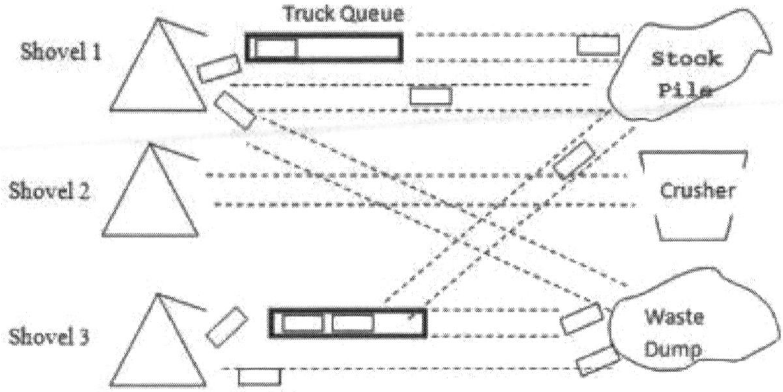

Figure 2—Schematic of truck and shovel operation

Previous studies (Burt *et al.,* 2005) have shown that different approaches, including deterministic, stochastic, and experimental methodologies, result in considerable differences in outputs. These techniques lead to different solutions regardless of the quality of the technique/software itself or the knowledge of the modelling team. Hence, the first step is to develop a comprehensive simulation framework for the problem of truck and shovel selection, sizing, and dispatching in open pit mines that obtains nearly the same optimal results for the same input variables, regardless of the technique employed (Burt, and Caccetta, 2014). In this study, a simulation technique was selected to solve the problem due to: i) possibility of incorporating uncertainties in different governing activities of the system, ii) extensive background of application of the technique in previous research and real world practices, iii) dynamic nature of the technique, which makes it applicable during the entire life of mine, and iv) relatively wide range of available software and languages.

This paper introduces a methodology and sensitivity analysis procedure for mine loading and haulage system selection and sizing. In addition, the capability of the DES technique in bulk material handling simulation is demonstrated through the application of the GPSS/H simulation language. The proposed framework was validated and tested in a large open pit mine. All steps in the proposed framework were followed attentively to ensure its effectiveness. However, due to space limitation, only the most important main components are discussed here, *i.e.* problem definition, data acquisition, statistical analysis, simulation language (technique) selection, model construction, model verification and validation, and sensitivity analysis. More information about the simulation history and GPSS/H background is given in Schriber (1992), Hollocks (2006), Nance (1995),

Pidd and Carvalho (2006), and Robinson, (2005). The paper is organized as follows:

> Proposed simulation frameworks for both new and existing systems

> Introduction of the case study

> Model building, verification, and validation

> Flow chart for performing sensitivity analysis

> Results, with a discussion of the optimal results

> Concluding remarks.

Several important steps of simulation of the case study are presented in more details in the Appendix.

Simulation framework

The lack of a comprehensive simulation framework in this field has resulted in considerably different solutions to the problem of truck and shovel system selection and sizing. Major source of these confusing differences include, but are not limited to:

> Different simulation approaches

> Different data requirements (quantity, quality, and statistical methodology)

> Insufficient technical communication during all phases of the project

> Insufficient determination of the objectives, resources, and constraints.

This study proposes a truck and shovel simulation framework for minimizing the errors due to erroneous or inaccurate assumptions and procedures, and provides a step-by-step simulation guideline. The algorithm attempts to render a framework for truck and shovel operation simulation. In the construction of the simulation framework, different blocks (Figure 2) were obtained from most of the available journal articles. The most important findings of the previous studies were selected and incorporated to achieve an efficient simulation strategy.

The first step in developing this framework, with the goals of completeness, comprehensiveness, and robustness, was to identify the major components of a general simulation modelling practice, regardless of the area of application. A general simulation framework is illustrated in Figure 3. This primary platform was set to serve as the structure of the framework and consequently was customized through the introduction of open pit mining specifications. These specified characteristics were derived from published articles in the field of mining operations simulation and modelling and were incorporated in the base structure. The base framework was composed of the following components:

> Problem definition, objectives, resources, and limitations:

> Data acquisition and statistical processing

> Model construction

> Model modification, verification, and validation

> Sensitivity analysis and decision-making strategies.

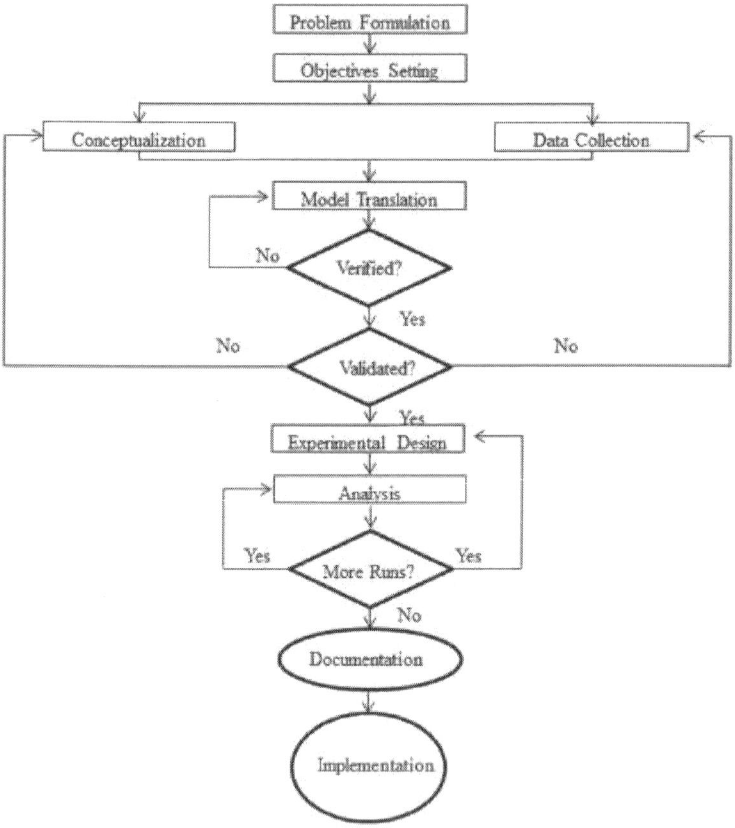

Figure 3—A general simulation flow chart (Banks, 2010)

It should be noted that there are several pitfalls in a general simulation practice (Maria, 1997) as follows: unclear objective, invalid model, simulation model too complex or too simple, erroneous assumptions,

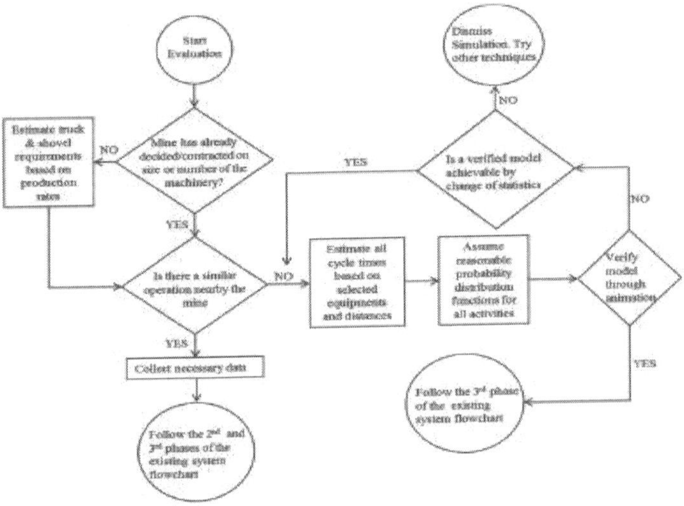

Figure 4—The proposed simulation framework for new mines

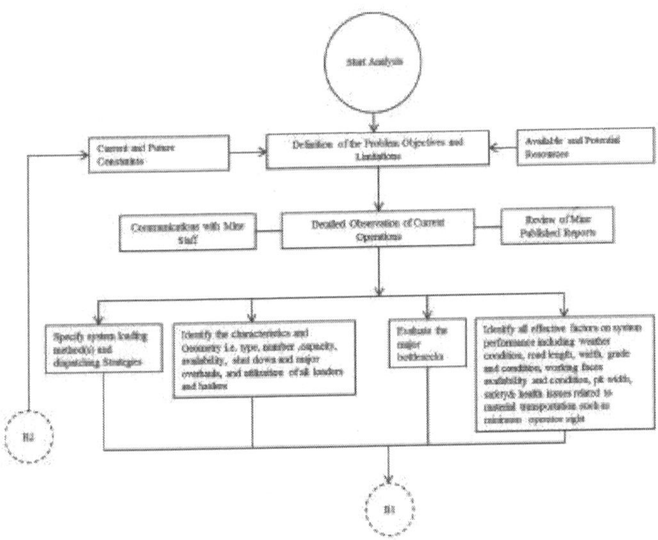

Figure 5a—The proposed simulation framework for existing mines; Phase I: Preparation and Information

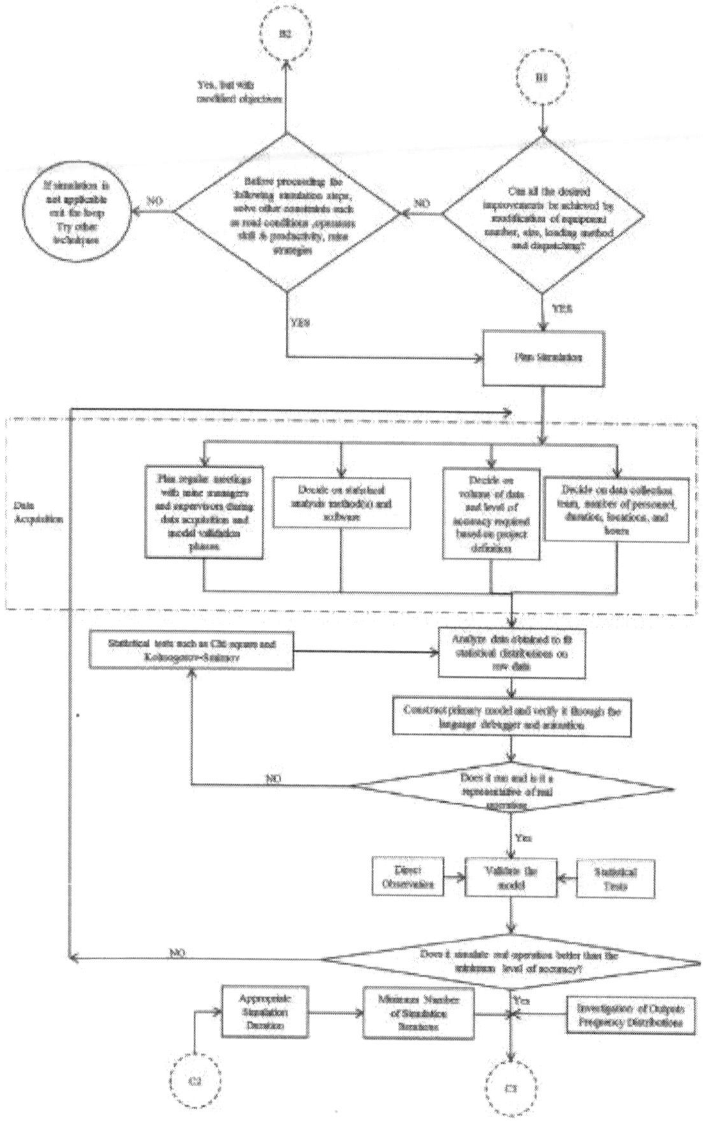

Figure 5b—The proposed simulation framework for existing mines; Phase II: Execution

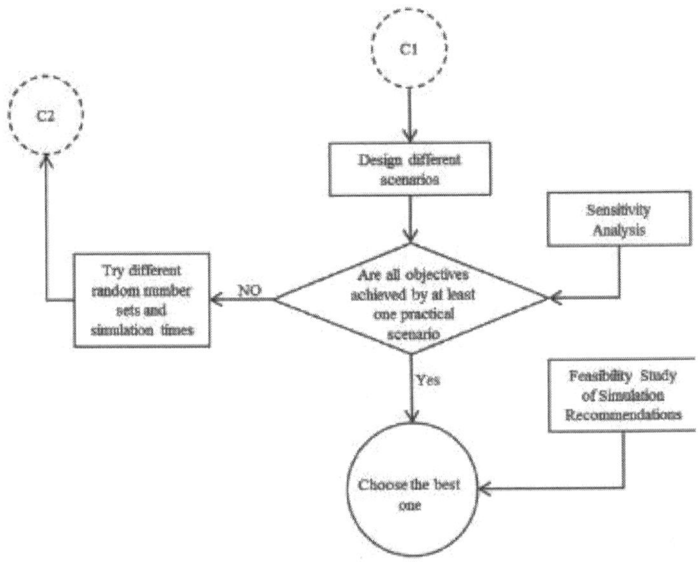

Figure 5c—The proposed simulation framework for existing mines; Phase III: Evaluation and decision

undocumented assumptions, and using the wrong input probability distribution.

The above pitfalls were incorporated in the proposed simulation framework for the truck and shovel selection and sizing problem.

The secondary mine-specific characteristics that contribute to mining operational performance include:

(1) Incorporation of the mining-environmental induced constraints

(2) Different traffic-dispatching scenarios

(3) Different loading methods

(4) Selection of hybrid or uniform loading/haulage fleets.

The framework is divided in two categories, for new and existing systems (Figures 4 and 5 respectively). Since in a new mine there is no operational data available, the simulation procedure needs extra considerations. These considerations are illustrated in the flow chart of Figure 4. In addition, a sensitivity analysis algorithm is presented later (Figure 12) that follows the simulation framework to evaluate different scenarios in mine truck and shovel system selection and sizing.

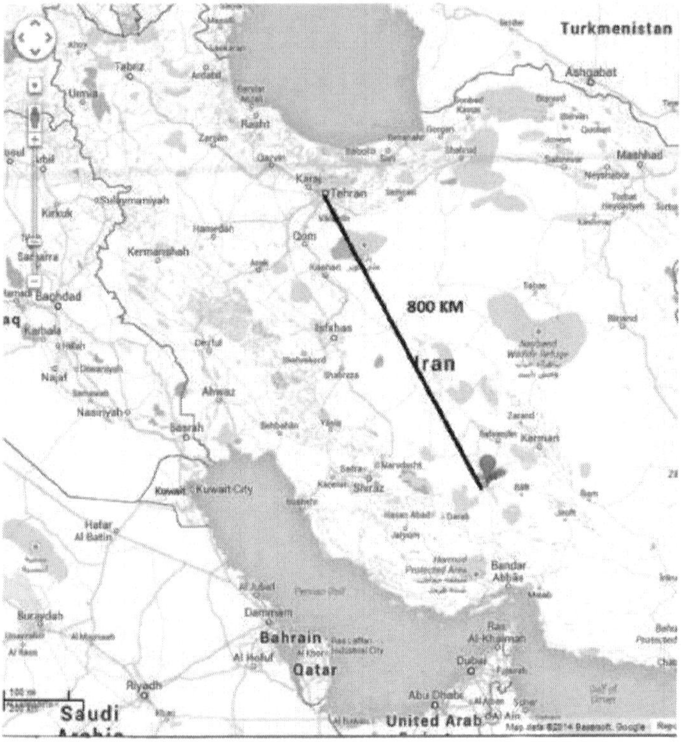

Figure 6—Location of Golegohar iron ore mine

Figure 7—Working area showing main ramps, roads and dumps (source: Google Earth, 2014)

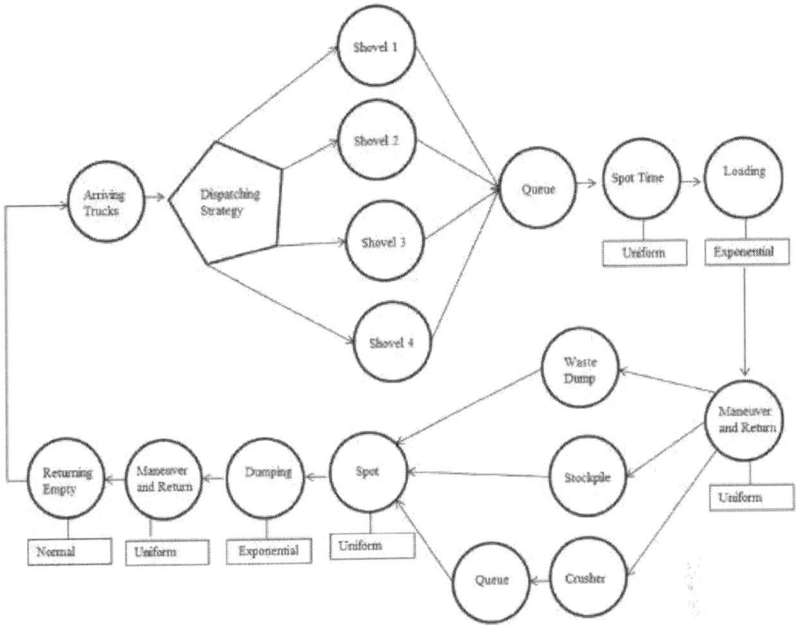

Figure 8—Block diagram of the case study operation flow

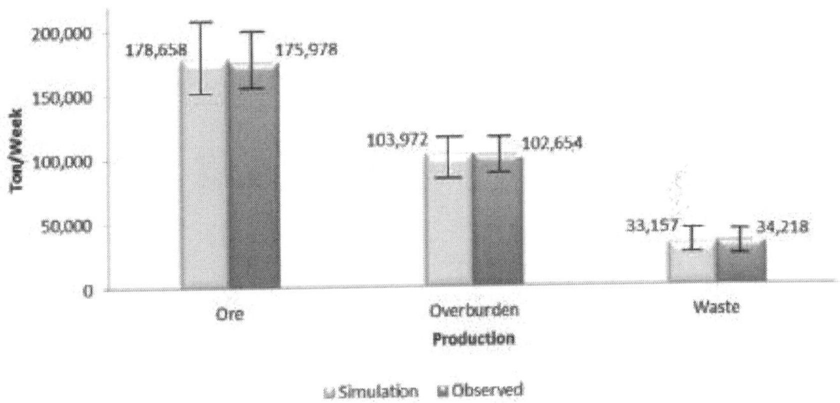

Figure 9—Production comparison

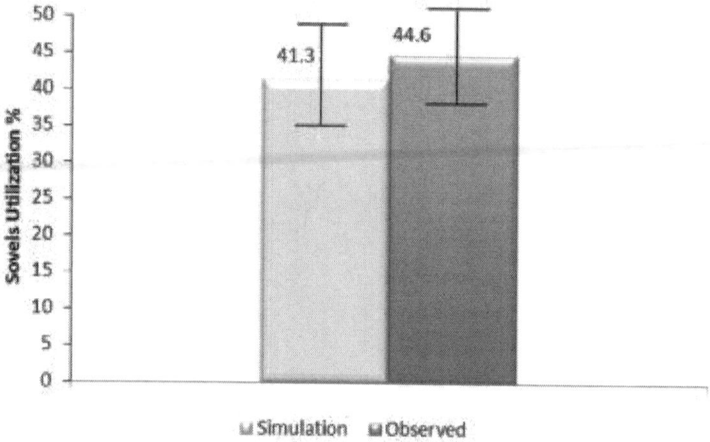

Figure 10—Shovel utilization

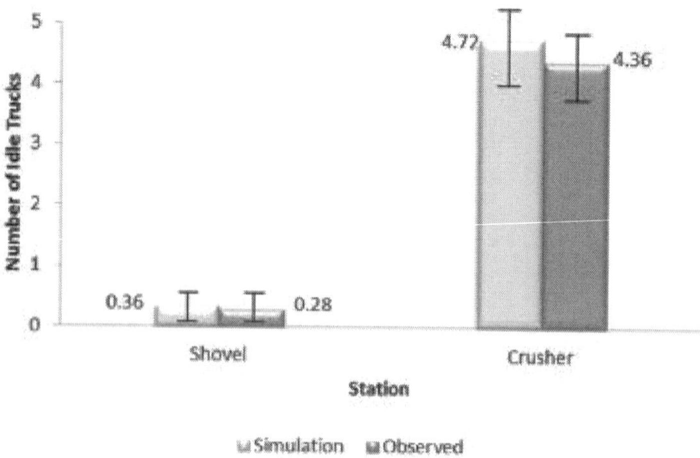

Figure 11—Number of idle trucks (queue length)

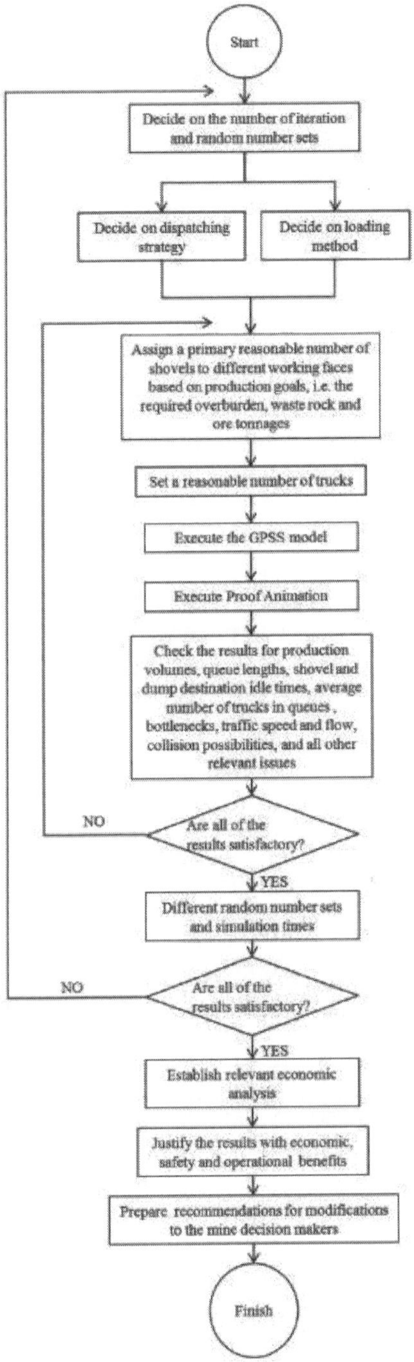

Figure 12—Flow chart of the sensitivity analysis

The main advantage of this simulation framework in comparison with other research is its comprehensive addressing of the problem of truck and shovel selection. All other available practices try to find solutions to specific parts of the problem, mainly in the form of what-if analysis. For instance, what would be the effect of adding one extra truck to the haulage fleet? Moreover, the framework is capable of addressing both a new and an existing open pit mine operation. This framework can add to the strength of simulation techniques in solving the problem compared to other abovementioned methods, which address the problem only partially.

Although this framework was validated in a large surface mine in this study, for other projects some modifications should be incorporated accordingly. For instance, production planning strategies in a mine with restricted processing plant requirements or ore grade limits dictate more frequent relocations of working faces, compared with a mine with more stable and predictable ore grade fluctuations. These differences introduce frequent changes in haulage distances and, hence, to the simulation approach at hand. Another example is the difference between a small surface mine with more short-term concentrated production plans and a large mine with more strategic and long-term plans. These types of specifications require more or less consideration of some blocks of the framework than others, accordingly (Figure 5).

CASE STUDY

Golegohar iron ore mine is located in southern Iran, 50 km from Sirjan, in the southwest of Kerman Province (latitude 29°7'N and longitude 55°19'E, Figure 6). This iron complex includes six known ore reserves and is one of the largest producers and exporters of iron concentrate in the country. It has a measured and indicated reserve of over 1 100 Mt of ore (Golegohar Iron Ore Complex, 2006). In Golegohar, over 10 Mt of iron concentrate is produced annually, through crushing, dry and wet grinding, and low-intensity magnetic separation (Figure 7).

To test the simulation framework, the operation of the current haulage system at Golegohar was investigated and necessary data collected. After statistical analysis of the raw data and deriving probabilistic distributions for each data-set (Carlsberg, 2011), a DES model was coded in GPSS/H (student version). In the next step, the proposed model was verified and then validated by comparing it with the real mine data by means of statistical tests *(e.g.* chi square) (Zimmermann, 2008). A series of sensitivity analyses were performed for the purpose of establishing the optimum number of cable shovels and dump trucks required to meet the production targets with the maximum possible system productivity. The simulation model was run with different combinations of the truck and shovel system numbers in a matrix pattern in order to identify the most appropriate system. The major constraint was set to achieve the annual production rate, and the main goal to accomplish the job with a minimum amount of equipment. Different scenarios were tested to find the highest equipment utilization and minimum idle and waiting time in queues. A dispatching system was introduced with the main objective of minimizing shovel idle time and the number of trucks in queues. The most important data acquired in the observation phase included mine production plans and layouts, current fleet geometry and parameters, and time data of the real system. Details of the applied methodology are illustrated in Figure 5.

Observation of the current system

Direct observation of the loading and haulage operations over 150 days, during different shifts and hours, resulted in recognition of the following problems in the current system:

> Truck loading method was single-sided

> On many occasions, only two out of the four cable shovels were operating. The other two were not used, either because of frequent mechanical failures or lack of proper working faces

> Shovels idle times were too long (mostly over 6 minutes). On some occasions, three or more trucks were arriving together (trucks queue)

> Almost always, there was a long queue before the primary crusher. The average queue length was four trucks (in random observations) and the average waiting time for each truck was 11.5 minutes

> There was no haulage fleet dispatching system in the mine

> The main bottleneck in the haulage fleet flow was at the primary crusher.

Model construction

After obtaining and analysing all necessary data, the next step was to construct a simulation model by programming in GPSS/H (student version). GPSS/H (General Purpose Simulation System) is both a computer language and a computer program. It was designed for studying systems represented by a series of discrete events. GPSS/H is a low-level, nonprocedural language. GPSS/H was selected as the simulation language as part of this study, for the following reasons:

(i) It is multivendor, so it is continually being upgraded

(ii) It is widely available

(iii) It is written in machine language and, therefore, is inherently very fast

(iv) It can solve a wide variety of problems rapidly and accurately

(v) It has proved to be extremely versatile for modelling mining operations. These include both surface and underground operations, as well as material flow through a smelter, mill, and refinery. It is easily coupled with PROOF for making animations (Sturgul, 2000).

A block diagram of the case study is illustrated in Figure 8.

It should be noted that, due to the application of probability distribution and random numbers in a queue system simulation (Figure 8), and for error reduction purposes, the final model, which consisted of 115 GPSS blocks, was executed with different random data-sets.

Model animation and validation

Some issues can be investigated easily in an animation that would be very hard to catch in the simulation model - like collisions or subtle logic glitches. The ability to see a model in action makes animation a great verification tool for the model

builder (Ståhl *et al,* 2011; Wolverine Software, 2013; University of Nevada, 2013). An animation model of the case study was executed by importing the GPSS/H model outputs to Proof Animation software. Observation of the animation model for different durations demonstrated that the proposed model logic had been achieved.

Model validation, as the most important phase of a DES exercise, was performed through comparison of the model outputs with the real system's data by designing statistical tests, *e.g.* chi-square and Kolmogorov-Smirnov at 5% significance level (Ross, 2006; Zimmermann, 2008). For this purpose, a new set of actual data, separate from the data used in the model, was collected. Three samples of the comparison results are illustrated in Figures 9-11, which were validated through chi-square method at the 5% significance level.

Further processing

A flow chart of the sensitivity analysis is presented in Figure 12. To evaluate the effectiveness of employing an appropriate traffic-dispatching strategy, the current model was modified to take this issue into account. The main goal of the dispatching algorithm was set to assign the incoming trucks first to the idlest shovels. However, many other what-if type questions may be answered by this model with minimum cost, safety issues, and disturbance to the current operation routines *e.g.* the effect of operators' skills, changes in road grades, the possibility of increasing the primary crusher's capacity, purchasing new trucks to replace some older ones with low mechanical availabilities (Burt *et al,* 2011), changing the current truck sizes (Bozorgebrahimi, Hall, and Morin, 2005), employing a hybrid haulage fleet, different dispatching strategies (Alarie and Gamache, 2002), feasibility of changing the current system to a conveyor belt system as the depth of the mine increases (Mcnearny and Nie, 2000).

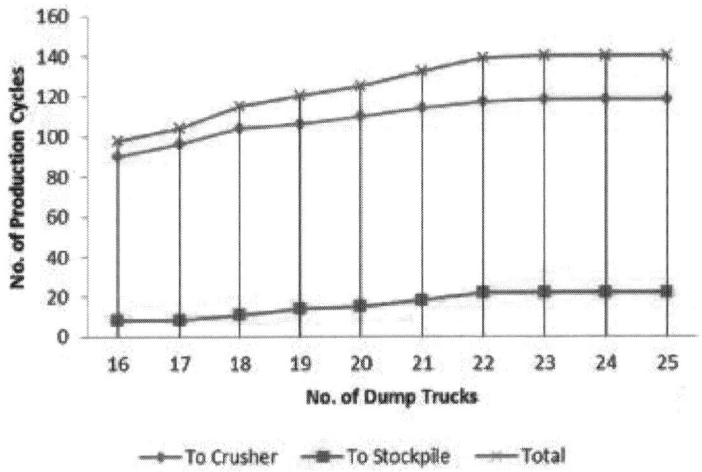

Figure 13—Effect of number of trucks on production (per simulated shift) for two cable shovels

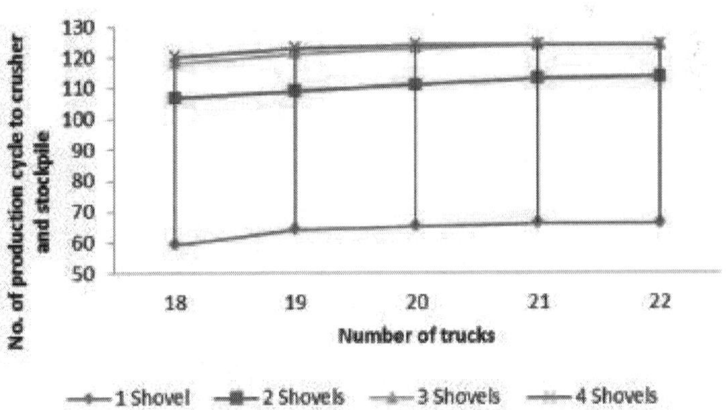

Figure14—Effect of both the number of trucks and cable shovels on ore production (per simulated shift)

The effect of the size of both the haulage and the loading fleet on effective shovel working hours is illustrated inFigure 15.

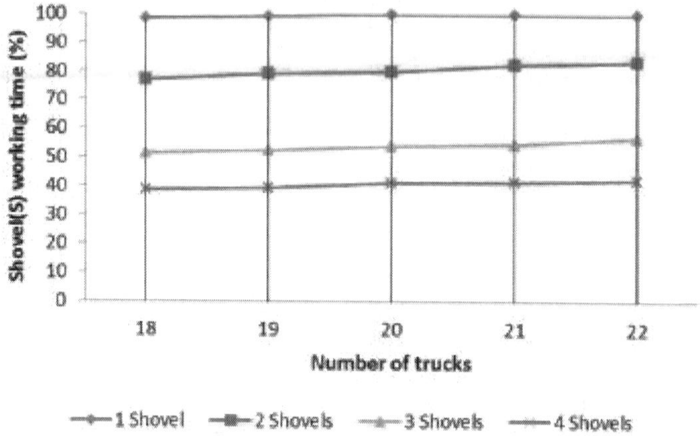

Figure 15—Effect of both the number of trucks and cable shovels on shovels working time % (per simulated shift)

Figure 16 shows the effect of application of a traffic dispatching system on the mine's operation. Regardless of the number of trucks, a proper dispatching strategy increases the productivity of the system.

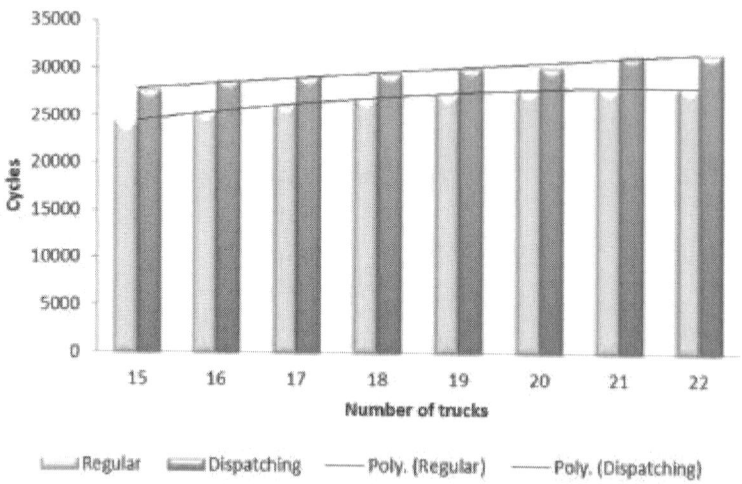

Figure 16—Regular vs. Dispatching (3-month plan)

Discussion of the simulation results

To demonstrate the capabilities of simulation in what-if analysis, a sample of the many analyses conducted in this study is discussed below:

Figure 13 illustrates the effect of the haulage fleet size on the number of production cycles per shift when only two shovels are operating. Increasing the number of dump trucks results in an increase of production rate per shift up to an optimum point. After that, because of the haulage fleet oversize and limited number of shovels (only two in this case), much time will be wasted in different truck queues at shovels. This results in no further increase in production. After this point the extra dump trucks in the system will be put into queues, thus making no contribution to production (Sturgul, 1995). Thus, the maximum production rate with two shovels is achieved with 22 dump trucks. This rate is 9.5 M t/a, which does not meet the minimum production requirements of the mine. At least three shovels are needed. The effect of the number of shovels on production rate is demonstrated in Figure 14. The simulation outputs show that increasing the number of shovels will result in more production with an equal number of trucks. For this range of truck numbers, operation with three and four shovels results in almost the same production rates. This is due to the fact that at least one of the shovels would be idle for most of a shift. Obviously, operating four loaders is not justified. At this point, three cable shovels will be the best choice for the mine's production target.

Summary of the optimal results

Table I shows that the modifications suggested by the simulation result in a 10% increase in production. Furthermore, using one less cable shovel contributes to increasing the mine profitability by introducing lower operational costs (assuming that the operational cost of one large shovel outweighs the associated costs of operating two additional trucks).

CONCLUSIONS

The truck and shovel system is the dominant method of material loading and haulage in open pit mining. Proper selection and sizing of the equipment has considerable effects on a mine's productivity and economics. A critical survey and analysis of the literature showed that deterministic, stochastic, and experimental methodologies for optimizing truck and shovel systems result in considerably different outputs. Thus, designing a comprehensive modelling framework is of high importance in system selection and sizing for mining operations. In addressing this issue, a stochastic simulation framework for truck and shovel system selection and sizing, for both new and existing open pit mines, was proposed. As part of the study, a proper simulation technique (discrete-event system simulation) and language (GPSS/H) were employed. Simulations were validated through real operations at a large open pit mine. The proposed framework is a useful guideline and should be applied accordingly based on the specific characteristics of the particular loading and haulage operation. Consideration of all the effective parameters and their interactions with the system, which are elaborated in the proposed framework, should be the top priority of a mine simulation team. Application of the proposed methodology resulted in considerable improvements in loading and haulage operations at a large open pit mine, with production rate increasing by about 10%.

REFERENCES

1. Aghajani, A., Osanloo, M., and Akbarpour, M. 2007. Optimising the loading system of Gol-e-Gohar iron ore mine of Iran by genetic algorithm. Australasian Institute of Mining and Metallurgy Publication Series. Melbourne. pp. 21-215.

2. Alarie, S. and Gamache, M. 2002. Overview of solution strategies used in truck dispatching systems for open pit mines. International Journal of Surface Mining, Reclamation and Environment, vol. 16, no. 1. pp. 59-76.

3. Awuah-Offei, K., Temeng, V.A., and Al-Hassan, S. 2003. Predicting equipment requirements using SIMAN, a case study. Mining Technology, vol. 112. pp. A180-A184.

4. Ayac, Z.Z. 2007. A hybrid approach to machine-tool selection through AHP and simulation. International Journal of Production Research, vol. 45, no. 9. pp. 2029-2050.

5. Baffi, E.Y. and Ataeepur, M. 1996. Simulation of a truck- shovel system using Arena. Proceeding of the 26th International Symposium on the Application of Computers and Operations Research in the Mineral Industries (APCOM), Pennsylvania, USA. pp. 153-159.

6. Banks, C.M. 2010. Introduction to modeling and simulation. Modeling and Simulation Fundamentals: Theoretical Underpinnings and Practical Domains. Sokolowski, J.A. and Banks, C.M. (eds.). Wiley, Hoboken, NJ. pp. 1-24.

7. Bozorgebrahimi, A., Hall, R.A., and Morin, M.A. 2005. Equipment size effects on open pit mining performance.International Journal of Surface Mining, Reclamation and Environment, vol. 19, no. 1. pp. 41-56.

8. Burt, C., Caccetta, L., Hill, S., and Welgama, P. 2005. Models for mining equipment selection. MODSIM05 - International Congress on Modellingand Simulation: Advances and Applicationsfor Management and Decision Making, University of Melbourne, 12-15 December 2005. Zerger, A. and Argent, R.M. (eds.). pp. 1730-1736.

9. Burt, C., Caccetta, L., Welgama, P., and Fouché, L. 2011. Equipment selection with heterogeneous fleets for multiple-period schedules. Journal of the Operational Research Society, vol. 62, no. 8. pp. 1498-1509.

10. Burt, C.N. and Caccetta, L. 2014. Equipment selection for surface mining: a review. Interfaces, vol. 44, no. 2. pp. 143-162.

11. Camarena, E.A., Gracia, C., and Cabrera Sixto, J.M. 2004. A mixed integer linear programming machinery selection model for multifarm systems. Biosystems Engineering, vol. 87, no. 2. pp. 145-154.

12. Carlberg, C. 2011. Statistical analysis; Microsoft Excel 2010. Indianapolis,IN.

13. Carr, J.R. 2003. Simple random number generation. Computers and Geosciences, vol. 29, no. 10. pp. 1269-1275.

14. Edwards, D.J., Malekzadeh, H., and Yisa, S.B. 2001. A linear programming decision tool for selecting the optimum excavator. Structural Survey, vol. 19, no. 2. pp. 113-120.

15. Golegohar Iron Ore Complex. 2006. Golegohar production reports.

16. Haldar, A. and Mahaderan, S. 2000. Probability, Reliability and Statistical Methods in Engineering Design. John Wiley & Sons, New York. pp. 304.

17. Hollocks, B.W. 2006. Forty years of discrete-event simulation: a personal reflection. Journal of the Operational Research Society, vol. 57, no. 12. pp. 1383-1399.

18. Komljenovic, D., Paraszczak, J., and Fytas, K. 2004. Optimization of shovel- truck systems using the queuing theory. CIM Bulletin, vol. 97. p. 76.

19. Krause, A. and Musingwini, C. 2007. Modelling open pit shovel-truck systems using the Machine Repair Model.Journal of the Southern African Institute of Mining and Metallurgy, vol. 107. pp. 469-476.

20. Maria, A. 1997. Introduction to modeling and simulation. Proceedings of the Winter Simulation Conference Proceedings, Atlanta, GA, 7-10 December 1997. Andraótthir, A., Healy, K.J.,

Withers, D.H., and Nelson, B.L. (eds.). Association for Computing Machinery, New York. pp. 7-13.

21. Marzouk, M. and Moselhi, 0. 2003. Constraint-based genetic algorithm for earthmoving fleet selection. Canadian Journal of Civil Engineering, vol. 30, no. 4. pp. 673-683.

22. Mcnearny, R.L. and Nie, Z. 2000. Simulation of a conveyor belt network at an underground coal mine. Mineral Resources Engineering, vol. 9, no. 3. pp. 343--355.

23. Nance, R.E. 1995. Simulation programming languages: an abridged history. Proceedings of the 1995 Winter Simulation Conference, Arlington, VA. pp. 1307-1313.

24. Pidd, M. and Carvalho, A. 2006. Simulation software: not the same yesterday, today or forever. Journal of Simulation, vol. 1, no. 1. pp. 7-20.

25. Robinson, S. 2005. Discrete-event simulation: from the pioneers to the present, what next? Journal of the Operational Research Society, vol. 56, no. 6. pp. 619-629.

26. Ross, I., Casten, T., Marsh, D., and Peppin, C. 2010. The role of simulation in ground handling optimization at the Grasberg block cave mine. Hoist and Haul 2010 - Proceedings of the International Conference on Hoisting and Haulage, Las Vegas, 12-17 September 2010. Brokenshire, P. and Andersen, S. (eds.). Society for Mining, Metallurgy & Exploration, Littleton, CO. pp. 257-265.

27. Ross, S.M. 2006. Statistical Modeling and Decision Science: Simulation. 4th edn. Academic Press, Burlington, MA.

28. Runciman, N., Vagenas, N., and Newson, G. 1996. Simulation modeling of underground hard - rock mining operations using WITNESS. Proceedings of the 26th International Symposium on the Application of Computers and Operations Research in the Mineral Industries (APCOM), Pennsylvania, USA. pp. 148-151.

29. Schriber, T.J. 1991. An Introduction to Simulation using GPSS/H. John Wiley & Sons, New York.

30. Schriber, T.J. 1992. Perspectives on simulation using GPSS. Proceedings of the 24th Conference on Winter Simulation, Arlington, VA, 13-16 December 1992 . ACM Press. pp. 338-342.

31. Sogaard, H.T. and Sorensen, C.G. 2004. A model for optimal selection of machinery sizes within the farm machinery system. Biosystems Engineering, vol. 89, no. 1. pp. 13-28.

32. StAhl, I., Henriksen. J., Born, R., and Herper, H. 2011. GPSS 50 years old, but still young. Proceedings of the Winter Simulation

Conference, Phoenix, AZ, 11-14 December 2011. Jain, S. Creasey, R.R., Himmelspach, J., White, K.P., and Fu, M. (eds). Institute of Electrical and Electronics Engineers, New York. pp. 3947-3957.

33. Stout, C.E., Conrad, P.W., Todd, C.S., Rosenthal, S., KnudsenInt, HP. Simulation of a large multiple pit mining operation using GPSS/H. Journal of Mining and Mineral Engineering, 2013, vol. 4, no. 4. pp. 278-295.

34. Sturgul, J.R. 1992. Using exact statistical distributions for truck shovel simulation studies. International Journal of Surface Mining and Reclamation, vol. 6, no. 3. pp. 137-139.

35. Sturgul, J.R. 1995. Simulation and animation come of age in mining. Engineering and Mining Journal, vol. 6, no. 2. pp. 38-42.

36. Sturgul, J.R. 1999. Discrete mine systems simulation in the United States. International Journal of Surface Mining Reclamation and Environment, vol. 13. pp. 37-41.

37. Sturgul, J.R. 2000. Mine Design Examples Using Simulation. Society of Mining, Metallurgy and Exploration, Littleton, CO. .

38. Sturgul, J.R. and Thurgood, S.R. 1993. Simulation model for materials handling system for surface coal mine.Bulk Solids Handling, vol. 13, no. 4. pp. 817-820.

39. University of Nevada. 2013. Mine Systems Optimization and Simulation Laboratory. Mining and Metallurgical Engineering Department, University of Nevada, Reno. http://www.unr.edu/mining/research/mine-systems-optimization-and-simulation

40. Wolverine Software. 2013. http://www.wolverinesoftware.com.

41. Zimmermann, A. 2008. Stochastic Discrete Event Systems: Modeling, Evaluation, Applications. Springer.

3

STABILIZATION OF GOB-SIDE ENTRY WITH AN ARTIFICIAL SIDE FOR SUSTAINING MINING WORK

Hong-sheng Wang [1,*], Dong-sheng Zhang [2], Lang Liu [1], Wei-bin Guo [1], Gang-wei Fan [2], KI-IL Song [3,*] and Xu-feng Wang [2]

[1]School of Energy Engineering, Xi'an University of Science and Technology, Key Laboratory of Western Mine Exploitation and Hazard Prevention with Ministry of Education, Xi'an 710054, China

[2]School of Mines, State Key Laboratory of Coal Resources & Safe Mining, China University of Mining & Technology, Xuzhou 221116, China

[3]Department of Civil Engineering, Inha University, Incheon 402-751, Korea

ABSTRACT

A concrete artificial side (AS) is introduced to stabilize a gob-side entry (GSE). To evaluate the stability of the AS, a uniaxial compression failure experiment was conducted with large and small-scale specimens. The distribution characteristics of the shear stress were obtained from a numerical simulation. Based on the failure characteristics and the variation of the shear stress, a failure criterion was determined and implemented in the strengthening method for the artificial side. In an experimental test, the distribution pattern of the maximum shear stress showed an X shape, which contributed to the failure shape of the specimen. The shear stress distribution and failure shape are induced by a combination of two sets of shear stresses, which implies that failure of the AS follows the twin shear strength theory. The use of anchor bolts, bolts, and anchor bars enhances the shear strength of the artificial side. When this side is stable, the components can constrain the lateral deformation as well as improve the internal friction angle and cohesion. When the AS is damaged, the components prevent the sliding of broken blocks along the shear failure plane and improve the residual strength of the artificial side. When reinforced with an anchor bar, the AS is still stable even after mining operations for three years.

Keywords: gob-side entry; artificial side; failure criterion; twin-shear failure; uniaxial compression failure experiment

1. INTRODUCTION

A concrete artificial side (AS) at a gob-side entry (GSE) is an important part for GSE retaining. The long-term stability of the concrete AS is a key issue for successful mining practice [1,2]. However, the long-term stability of AS can be affected by the state of stress and deformation characteristics. The AS has to experience mining-induced influences and long-term creep deformation (1–2 years), which can easily lead to stress or deformation failure. If the stress and deformation cannot be controlled effectively, the AS will be destabilized, and the whole gob side entry will finally collapse. Therefore, studying the failure criteria and the stress variation of the AS is important for successful implementation of a GSE retaining. Based on a theoretical study, a failure criterion can be implemented in practical design, which will improve the recovery rate of coal resources and ensure working safety.

The behavior of a material can be defined with a strength criterion. The Tresca yield criterion [3] can be used to define the behavior of an elastic material. When the maximum shear stress reaches a certain value, the Tresca material will yield and slide along the maximum shear stress direction. In the Mohr criterion [4], the failure of a certain point in the material is mainly determined by the maximum principal stress and the minimum principal stress. The failure of a Mohr material is independent of the intermediate principal stress. The ultimate shear strength of the material is closely related to the cohesion and the internal friction angle in Coulomb criterion [5]. These three representative strength criteria only consider the influence of normal stress in a shear stress plane. Thus, these strength criteria consider only the maximum and minimum principal stress, but not the intermediate principal stress, and are known as single shear strength theory.

Yu (1983) proposed a generalized twin shear strength theory that is appropriate for a material that has different strength in tension and compression [6,7,8,9]. The failure mode of the AS is affected by many factors [10,11,12], especially dynamic load induced by blasting which can cause significant damage to AS. In this study, we determined the failure characteristics and the variation of shear stress of the concrete AS of the GSE, which were affected by the coal mining. To achieve this, a uniaxial compression failure experiment was conducted with large and small-scale specimens. The distribution characteristics of shear stress were obtained from a numerical simulation. Based on the results, a failure criterion was determined and implemented in a strengthening method for the artificial side.

2. STRUCTURAL MODEL OF THE ARTIFICIAL SIDE

A schematic diagram of a GSE is presented in Figure 1. The immediate roof above the AS prevents ground movement and rock deformation induced by GSE formation. The stiffness of the main roof and immediate roof are assumed to be much greater than that of the AS and the rigid floor that supports it. Thus, deformation occurs at the upper boundary of the artificial side, and the vertical displacement at the lower boundary is ignorable. Based on these conditions, a structural model of the AS was established, as shown inFigure 2.

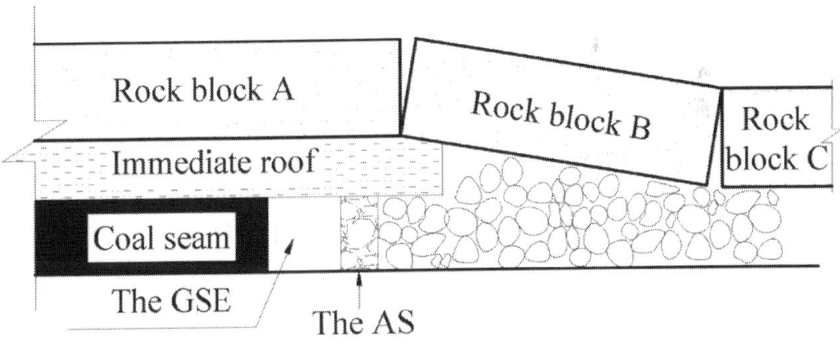

Figure 1. Schematic diagram of gob-side entry (GSE).

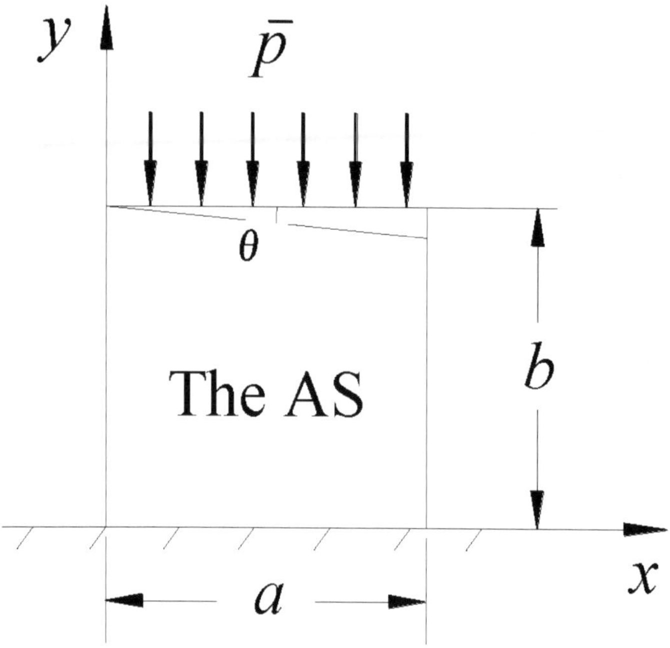

Figure 2. Structural model of the artificial side. a: width, b: height, θ: rotating angle of main roof, p: load.

During the entire service period, the AS experiences three stages [1,2]. The bearing load of the AS in the process of face mining can be classified into two parts: the dead weight of the immediate roof and main roof, and the additional load induced by rotation of the main roof [13]. Thus, high abutment pressure develops behind the working face. The deformation of the AS increases due to the abutment pressure. The fracture on the AS induced by abutment pressure decreases its structural integrity and load-carrying capacity. Finally, the failure of the AS can cause overall structural instability of the GSE.

3. FAILURE CRITERION OF ARTIFICIAL SIDE

3.1. Uniaxial Compression Failure Experiment

The mixture ratio of a specimen is determined according to the specifications for the mix proportion design of ordinary concrete (JGJ 55-2011) [14], as shown in Table 1. In the process of mixing, the amounts of the various materials need to be adjusted based on the moisture content of the sand and the particle size of the pebbles.

Table 1. Concrete mixture ratio scheme.

Strength (MPa)	Density of Materials (kg·m^{-3})				Water-Cement Ratio
	Water	Cement	Sand	Pebble	
10	180	230	780	1240	0.78

Based on Table 1, small and large specimens were fabricated. The small specimen was 70 mm × 70 mm × 70 mm, and the large specimen was 1500 mm × 600 mm × 900 mm (length × width × height, respectively). The specimens were moist cured to delay shrinkage for 28 days. One-time-concreting shaping technology was applied to cast the large specimen. The fabricated specimens are shown in Figure 3.

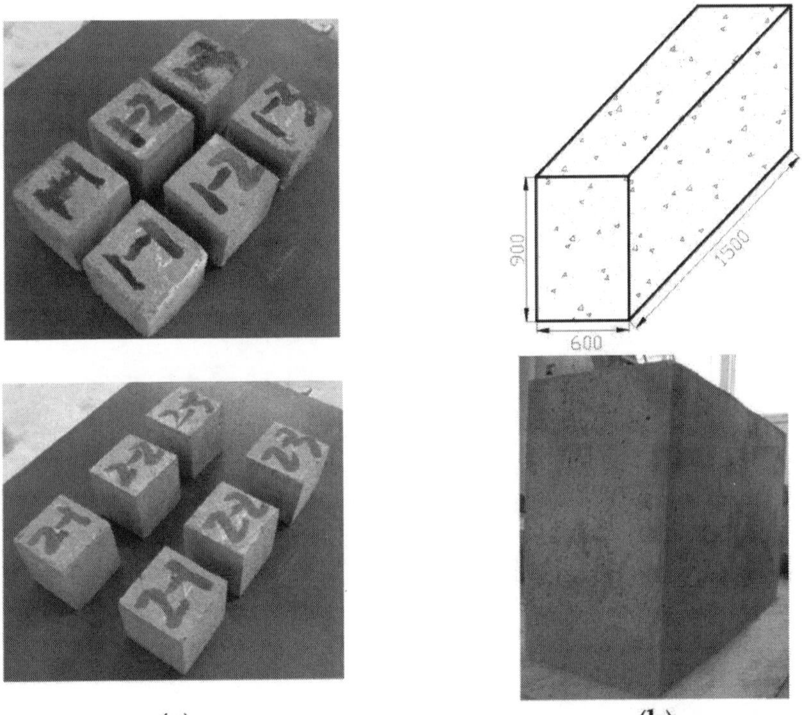

(a) (b)

Figure 3. Concrete specimens: (**a**) small-scale specimen; (**b**) large-scale specimen.

The uniaxial compression failure experiment was conducted with the custom-built large-scale experimental system shown in Figure 4 for the large-scale specimen. This system can be used to characterize the behavior of large-scale coal and rock under high loading conditions. The maximum size of a specimen is 1500 mm × 600 mm × 900 mm. The maximum pressure is 20 MPa, and the strain gauges have a strain measurement accuracy of one micro strain. For small-scale specimens, the experiment was conducted with a computer-controlled electronic universal testing machine.

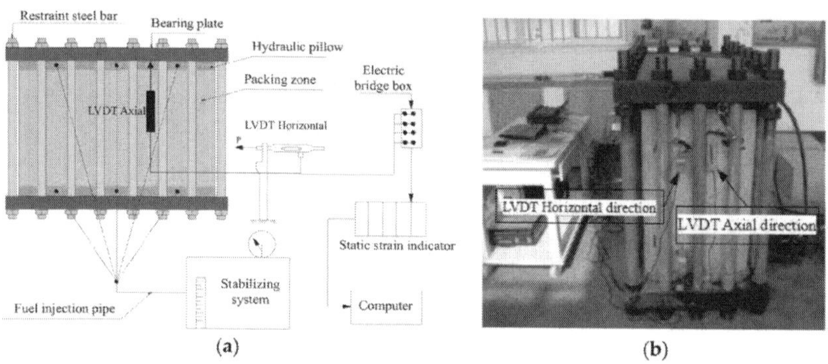

Figure 4. The uniaxial compression failure experiment system. (**a**) Schematic diagram of experiment system; (**b**) Material object of experiment system.

Figure 5 and Figure 6 present the failure pattern of a concrete AS for different specimen sizes. An X-shaped failure pattern was found in the small-scale specimens. The angle between the failure plane and the upper and lower planes ranged from 48° to 56°, as shown in Figure 5. The failure pattern of the large-scale specimen also appeared X-shaped, as shown in Figure 6. The AS showed brittle failure after transfixed fractures, and then it quickly lost its load-carrying capacity.

Figure 5. X-shaped failure pattern of small-scale specimen.

From Figure 5 and Figure 6, it can be seen that the failure pattern of the specimen appears X-shape regardless of specimen size. Unfortunately, it is hard to explain this failure pattern with one maximum shear stress (τ_{13}). It is clear that the X-shaped failure pattern results from two sets of shear stresses (τ_{13} and τ_{12}), as shown in Figure 7.

3.2. Distribution of the Maximum Shear Stress

An experiment can be conducted to characterize the failure mode. The variation and distribution of the maximum shear stress of a specimen cannot be monitored in the experiment, so numerical analysis was performed using the commercial nonlinear analysis software LS-DYNA. The Drucker-Prager criterion was adopted for the simulation of a specimen, and the mechanical parameters of constitutive model are presented in Table 2. The simulation process was divided into 10 load steps, thus the 2 MPa of step load was imposed on top of the specimen until the axial deformation reached 3%. The boundary condition for the numerical model was identical to the experimental set up, and the four sides of the specimen were free to deform.

Figure 8 shows the variations of the maximum shear stress of the small specimen during a uniaxial compression failure experiment. A shallow maximum shear stress band appeared at upper and lower parts of the specimen in the initial stage of loading. As the load increased, the shear stress level increased at the center of the specimen. The shear stresses that started from the upper and lower parts of the specimen overlapped and formed the X-shaped maximum shear stress distribution shown in Figure 8f. The distribution pattern of the maximum shear stress is basically consistent with the failure pattern of the small-scale specimen in Figure 5. It can be concluded that the distribution pattern of the maximum shear stress determines the failure pattern of the small-scale specimen.

Figure 6. Failure characteristics of large-scale specimen.

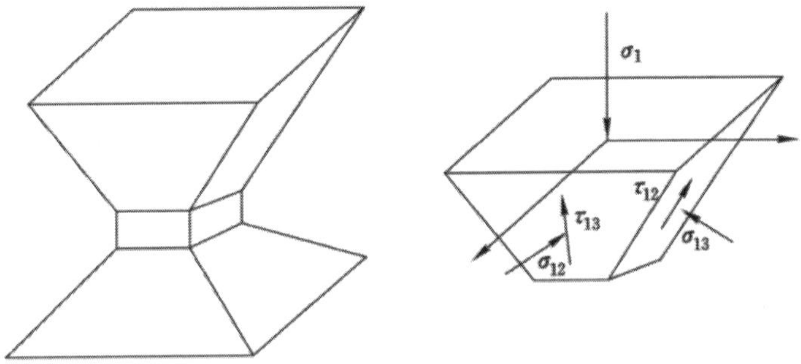

Figure 7. Failure characteristics of a test specimen.

Table 2. Mechanical parameters for numerical analysis.

Bulk Modulus (GPa)	Poisson Ratio	Shear Modulus (GPa)	Density (kN/m^3)	Internal Friction Angle (°)	Cohesion (MPa)
35	0.23	23	24	40	22

The variation of the maximum shear stress in the large-scale specimen during the uniaxial compression failure test is shown in Figure 9. The maximum shear stress initially appeared at upper and lower parts of the specimen. Two internal shear stress bands developed in the middle of the specimen, and the distribution range of the maximum shear stress band was extended to the peripheral part as the load increased. The maximum shear stress band formed an X-shaped pattern, as shown in Figure 9f. The distribution pattern of the maximum shear stress is consistent with the failure pattern of the large-scale specimen in Figure 6. It can be concluded that the distribution pattern of the maximum shear stress determines the failure pattern of the large-scale specimen as well.

Figure 10 shows the contours of the maximum shear stress and maximum principal stress of the large-scale specimen. Based on the stress distribution pattern, the specimen was divided into four parts as shown in Figure 11. The upper and lower parts in Figure 10 and Figure 11 show the compressive stress zone, while the left and right parts show the tensile stress zone. The locations of these zones are closely related to the aspect ratio of the specimen. The area of the compressive stress zone is inversely proportional to the aspect ratio, while the area of the tensile stress zone is proportional to the aspect ratio [1,2]. The distribution pattern of the maximum shear stress of the specimen appears X-shaped, and the maximum shear stresses in the four shear planes are identical in opposite directions.

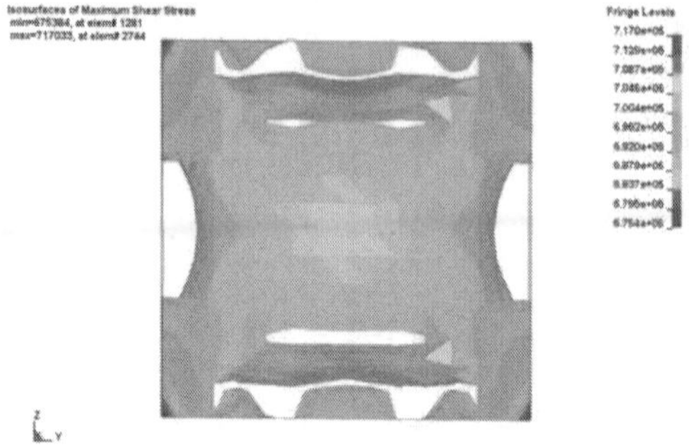
Isosurfaces of Maximum Shear Stress
Fringe Levels

(a)

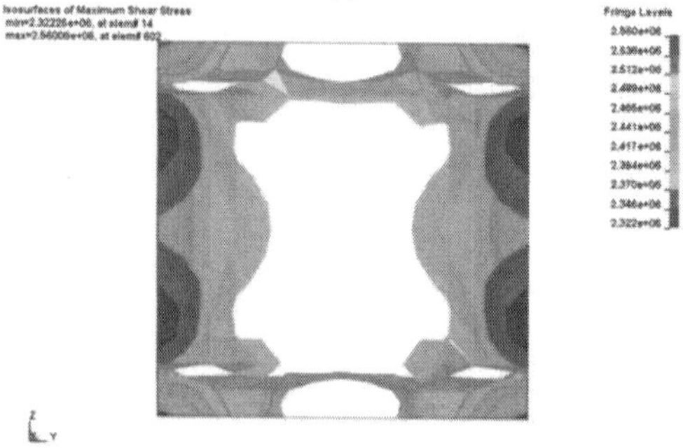
Isosurfaces of Maximum Shear Stress
Fringe Levels

(b)

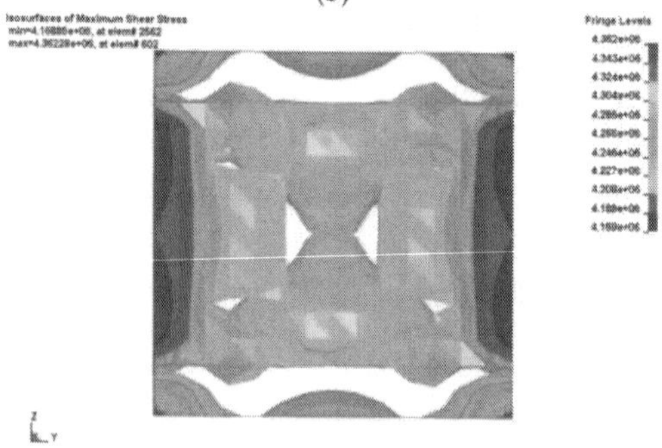
Isosurfaces of Maximum Shear Stress
Fringe Levels

(c)

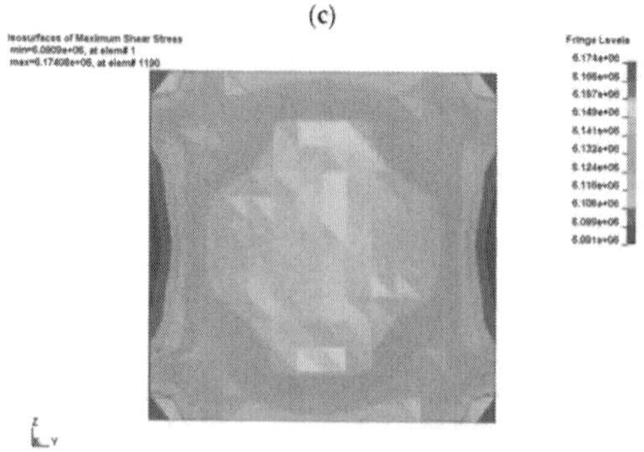
Isosurfaces of Maximum Shear Stress
Fringe Levels

(d)

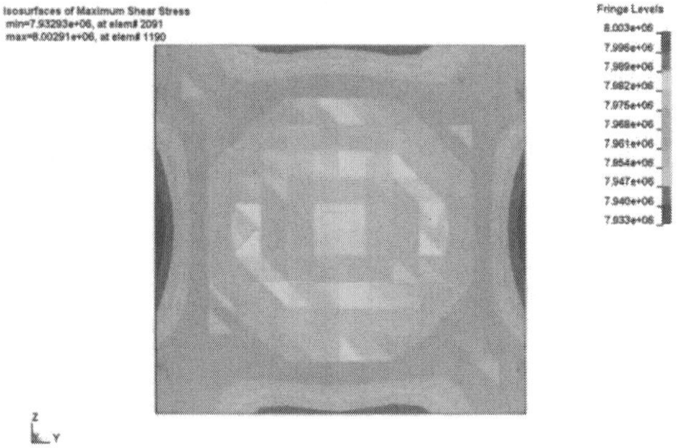

(e)

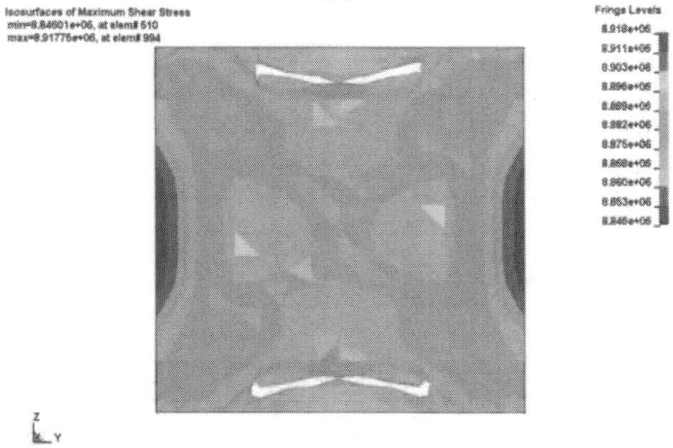

(f)

Figure 8. Variation of maximum shear stress of small-scale specimen. (**a**) loading step 1; (**b**) loading step 3; (**c**) loading step 5; (**d**) loading step 7; (**e**) loading step 9; (**f**) loading step 11.

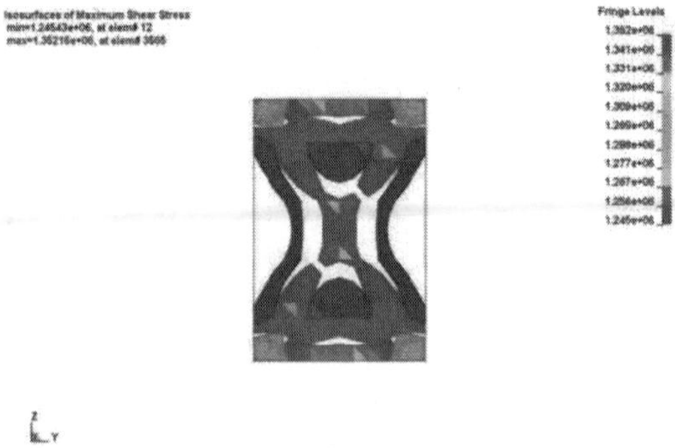
Isosurfaces of Maximum Shear Stress
Fringe Levels

(a)

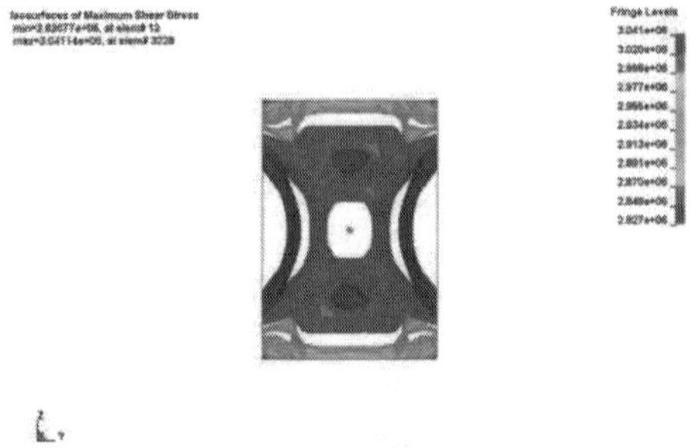
Isosurfaces of Maximum Shear Stress
Fringe Levels

(b)

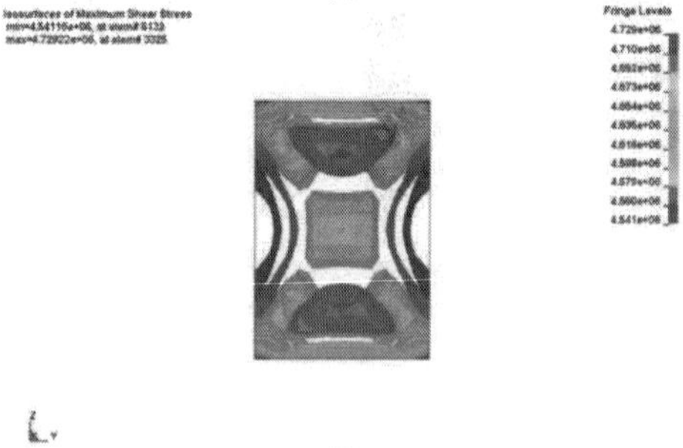
Isosurfaces of Maximum Shear Stress
Fringe Levels

(c)

Isosurfaces of Maximum Shear Stress
Fringe Levels

(d)

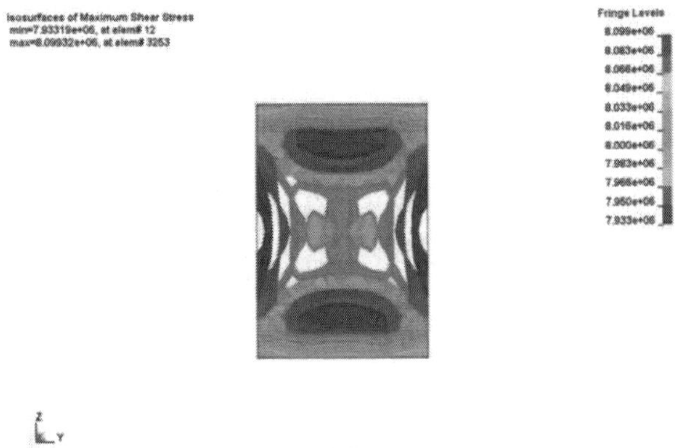

(e)

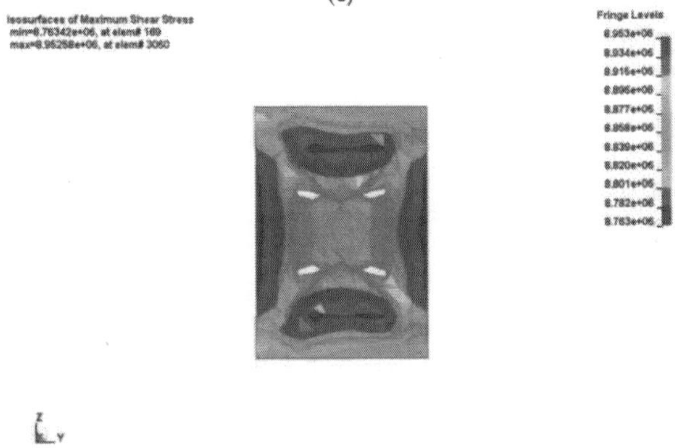

(f)

Figure 9. Variation of maximum shear stress of large-scale specimen. (**a**) loading step 1; (**b**) loading step 3; (**c**) loading step 5; (**d**) loading step 7; (**e**) loading step 9; (**f**) loading step 11.

3.3. Failure Criterion of the Artificial Side

3.3.1. Orthogonal Octahedron and Its Stress Function

The cube unit considered in this study is a space aliquot body. A cube unit is commonly used in general material mechanics and in elastic and plastic mechanics. It is composed of three pairs of mutually perpendicular sections, as shown in Figure 12a. When four sections inclined 45 degrees in σ_1 and σ_3 space intersect the cube unit, the space aliquot body becomes the unit body of maximum shear stress, as shown inFigure 12b. A set of mutually perpendicular sections with principal shear stress are used to cut the unit body of maximum shear stress, and the space aliquot body becomes the orthogonal octahedron shown in Figure 12c.

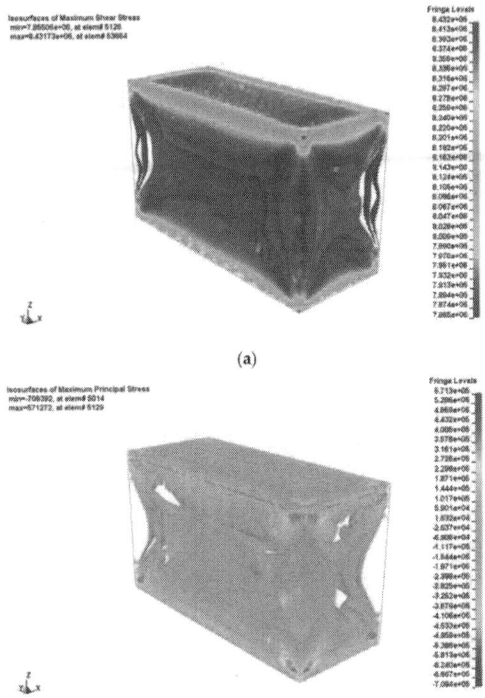

(a)

(b)

Figure 10. Stereogram of the iso-surface of large-scale specimen. (**a**) Maximum shear stress; (**b**) Maximum principal stress.

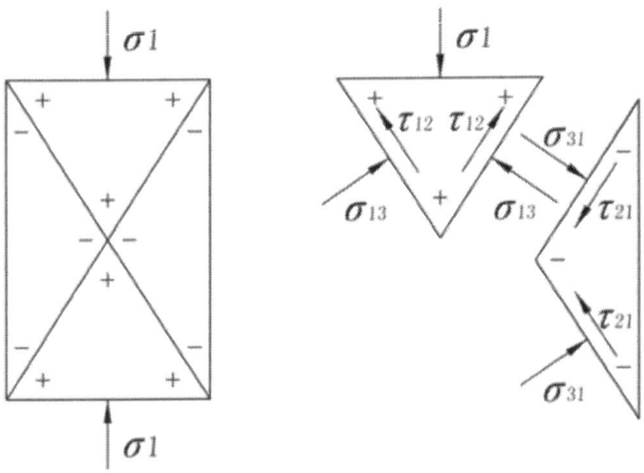

Figure 11. Stress analysis of the specimen.

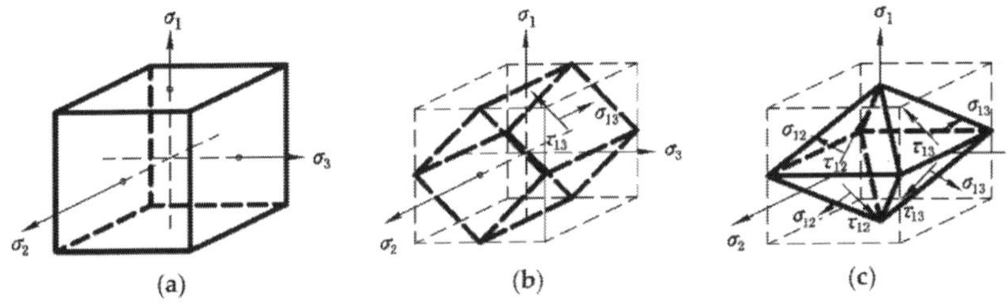

Figure 12. Space aliquot body. (a) Cube unit; (b) Unit body of maximum shear stress; (c) Orthogonal octahedron.

There are two sets of principal shear stresses in the orthogonal octahedron, τ_{13} and τ_{12}. Thus, it can be considered as a twin shear stress unit. The stress components in the orthogonal octahedron are τ_{12}, τ_{13}, τ_{23},σ_{12}, σ_{13}, and σ_{23}. The twin shear stress function can be defined as follows:

$$\begin{cases} T_\tau = \tau_{13} + \tau_{12} = \sigma_1 - \frac{1}{2}(\sigma_2 + \sigma_3) \\ T'_\tau = \tau_{13} + \tau_{23} = \frac{1}{2}(\sigma_1 + \sigma_2) - \sigma_3 \end{cases} \quad (1)$$

In this function, two straight inclined lines show the effect of intermediate principal stress σ_2.

3.3.2. Twin Shear Strength Theory

For the twin-shear stress unit shown in Figure 12c, the stress components can be defined as follows:

$$\begin{aligned} \tau_{12} &= \frac{1}{2}(\sigma_1 - \sigma_2) \\ \tau_{13} &= \frac{1}{2}(\sigma_1 - \sigma_3) \\ \tau_{23} &= \frac{1}{2}(\sigma_2 - \sigma_3) \end{aligned} \quad (2)$$

$$\begin{aligned} \sigma_{12} &= \frac{1}{2}(\sigma_1 + \sigma_2) \\ \sigma_{13} &= \frac{1}{2}(\sigma_1 + \sigma_3) \\ \sigma_{23} &= \frac{1}{2}(\sigma_2 + \sigma_3) \end{aligned} \quad (3)$$

The twin shear strength theory was established based on the concept of the twin shear stress state in the twin-shear stress unit [8]. In twin shear strength theory, the material will fail or yield when a critical state is reached in the influence function of the two sets of shear stress and normal stress on corresponding planes in the twin shear stress unit. The mathematical expression is defined as follows:

$$\begin{cases} F = \tau_{13} + \tau_{12} + \beta(\sigma_{13} + \sigma_{12}) = C\ , & (\tau_{12} + \beta\sigma_{12} \geqslant \tau_{23} + \beta\sigma_{23}) \\ F' = \tau_{13} + \tau_{23} + \beta(\sigma_{13} + \sigma_{23}) = C\ , & (\tau_{12} + \beta\sigma_{12} \leqslant \tau_{23} + \beta\sigma_{23}) \end{cases} \tag{4}$$

where β and C are material parameters that can be determined by the limit (σ_t, σ_c) of the tensile strength, compressive strength, and the ratio of the tensile strength to compressive strength ($\alpha = \sigma_t/\sigma_c$):

$$\beta = \frac{\sigma_c - \sigma_t}{\sigma_c + \sigma_t} = \frac{1-\alpha}{1+\alpha}, C = \frac{2}{1+\alpha}\sigma_t \tag{5}$$

By substituting Equations (2), (3), and (5) into (4), the principal stress in the twin shear strength theory can be written as follows:

$$\begin{cases} F = \sigma_1 - \frac{\alpha}{2}(\sigma_2 + \sigma_3) = \sigma_t\ , & \left(\sigma_2 \leqslant \frac{\sigma_1 + \alpha\sigma_3}{1+\alpha}\right) \\ F' = \frac{1}{2}(\sigma_1 + \sigma_2) - \alpha\sigma_3 = \sigma_t\ , & \left(\sigma_2 \geqslant \frac{\sigma_1 + \alpha\sigma_3}{1+\alpha}\right) \end{cases} \tag{6}$$

As shown in Equation (6), the twin shear strength theory reflects the influence of the intermediate principal stress σ_2 on the material behavior.

4. STRENGTHENING THE STABILITY OF ARTIFICIAL SIDE

4.1. Reinforcement of Artificial Side

According to the experimental and numerical studies, an X-shaped failure pattern is commonly found in an artificial side, regardless of specimen size. From a theoretical point of view, the X-shaped failure pattern is mainly induced by the combination of two sets of shear stresses. To enhance the performance of the artificial side, bolt-type reinforcements such as an anchor bolt, bolt, or anchor bar is suggested in this study as shown in Figure 13.

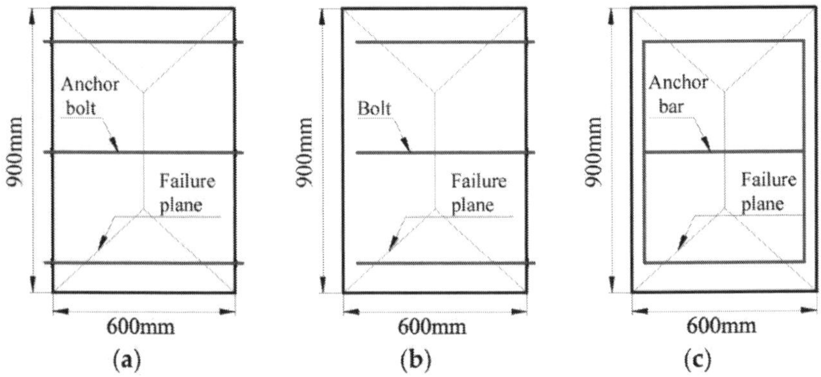

Figure 13. Bolt-type reinforcements of the artificial side. (**a**) Anchor bolt; (**b**) Bolt; (**c**) Anchor bar.

4.2. Reinforcement Mechanism

The limit equilibrium condition of the plastic softening material can be defined as follows:

$$\sigma_1 = \sigma_3 \frac{1+\sin\varphi}{1-\sin\varphi} + 2c\frac{\cos\varphi}{1-\sin\varphi} \tag{7}$$

where φ and c are the internal friction angle and the cohesion, and σ_1 and σ_3 are equivalent to the load-carrying capacity and lateral confining stress, respectively. The bolt-type reinforcement can develop a lateral confining effect, $\sigma_3 > 0$. From Equation (7), σ_1 increases as σ_3 increases. Increasing the lateral confining stress changes the state of stress on both sides of the AS from two-dimensional to three-dimensional. In the three-dimensional state of stress, the plastic property of the AS can be fully activated under excessive loading.

When the AS is stable, bolt-type reinforcement constrains the lateral deformation and improves the internal friction angle φ, cohesion c, and enhances the shear strength of the artificial side. When the AS is unstable due to damage, the constraining effect induced by bolt-type reinforcement can prevent a broken block from sliding along the shear failure plane and improve the residual strength of the artificial side. Thus, the load-carrying capacity of the AS can be sustained. Therefore, the bolt-type reinforcement improves φ and c of the AS and enhances the load-carrying and anti-deformation capacity.

4.3. Validation of Bolt-Type Reinforcement

An additional numerical study was performed to validate the effect of the reinforcement. It was assumed that nine anchor bolts were installed in a large-scale specimen. The variations of the maximum shear stress and maximum principal stress of the reinforced large-scale specimen are presented in Figure 14.

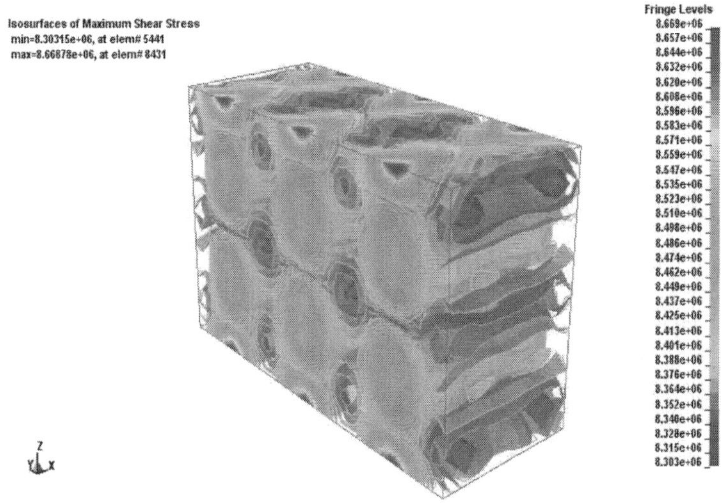

(a)

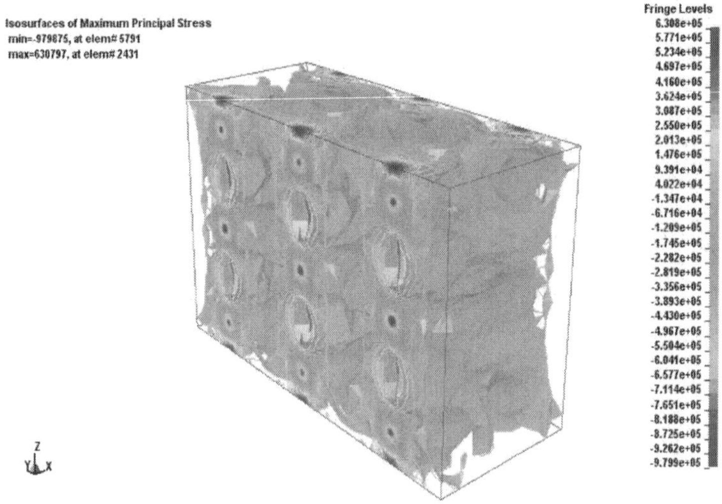

(b)

Figure 14. Stereogram of the stress distribution in reinforced large-scale specimen. (**a**) Maximum shear stress; (**b**) Maximum principal stress.

In Figure 14a, the distribution of the maximum shear stress changed: the X-shaped pattern disappeared, the maximum shear stress decreased, and the shear strength increased. The lateral deformation was also constrained significantly. In Figure 14b, the distribution pattern of the principal stress has changed: the load-bearing area expanded while the tensile area shrank. The maximum principal stress and the maximum tensile stress decreased. Due to the installation of anchor bar, the compressive strength of the AS increased, and the load-carrying capacity of the AS was improved.

5. APPLICATION OF ARTIFICIAL SIDE

5.1. Geological Conditions

Figure 15 shows the layout of the Left No. 1 Working Face in the West No. 2 Working Area of the Jixian Coalmine. The seam thickness of No. 9 ranged from 1.4–1.68 m with an average of 1.6 m. The seam pitch was 10°. The immediate roof was grey medium sandstone with a thickness of 1.7 m. The immediate floor was grey fine sandstone with a thickness of 0.6 m. The main roof and floor were black siltstone with thicknesses of 2.6 m and 3.3 m, respectively. A belt transportation roadway was tunneled along the roof of the No. 9 seam and used for air intake and coal transportation. The length of the roadway was 768 m. GSE retaining technology was adopted to increase the resource recovery rate. The section of the GSE is shown in Figure 16.

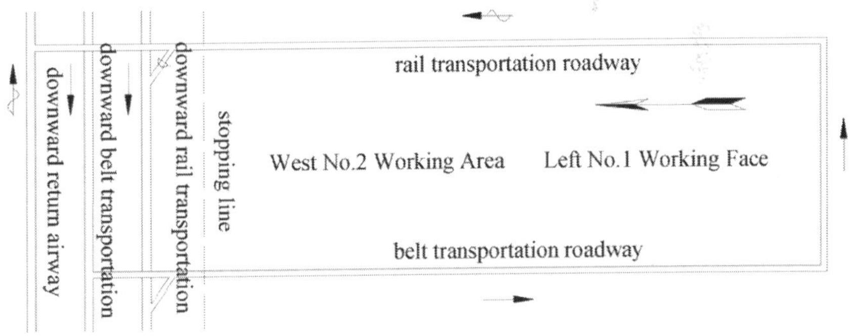

Figure 15. Layout of the working face.

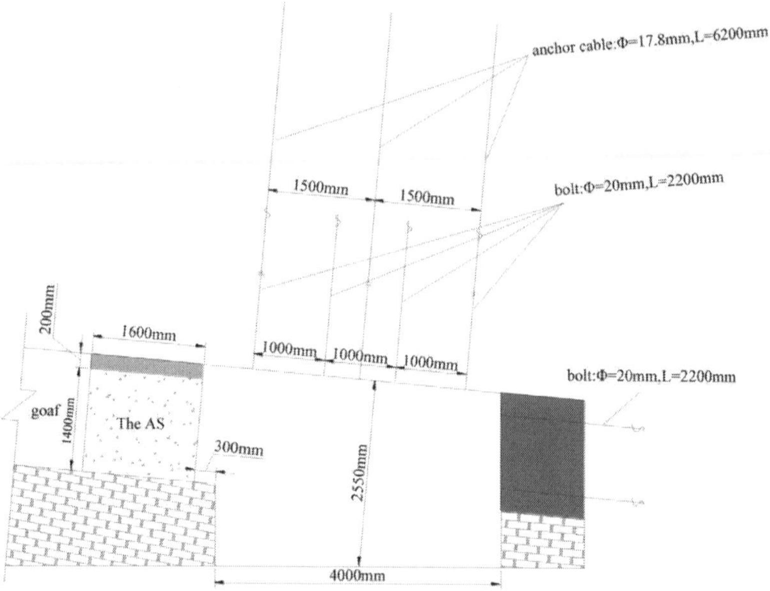

Figure 16. Section of the GSE.

5.2. Parameters of the Artificial Side

To decrease the load on the AS when the main roof fractures and rotates, a soft-hard structure was used for the AS [1,2,15]. Based on the seam thickness of the coal, the given deformation of the key block in the main roof was about 240 mm, so the thicknesses of the upper soft structure and the hard structure at the bottom were 240 mm and 1400 mm, respectively. The height of the AS was 1640 mm. The results [1,2] show that the load-bearing area and load-carrying capacity are highest when the aspect ratio is 1:1. Therefore, the width of the AS is decided as 1600 mm.

5.3. Construction Method of the Artificial Side

Considering the geological conditions of the working face, rail transport and monorail system were used to transport large concrete blocks that were fabricated in advance. Concrete blocks were stacked in two layers to build up the artificial side.

5.4. Reinforcement of Artificial Side

Anchor bars were installed in the AS to improve the load-carrying capacity. The anchor bar was thread steel, its diameter was 18 mm, and its length was 1600 mm. One hundred millimeters of both ends were folded inside, and the anchor bars were fixed with round steel of 14 mm in diameter. The anchor bar framework was fixed at the center of a mold, and then concrete paste was added. The AS is wet-cured for more than 28 days at normal temperature. The anchor bar framework and the large concrete blocks are shown in Figure 17a,b, respectively.

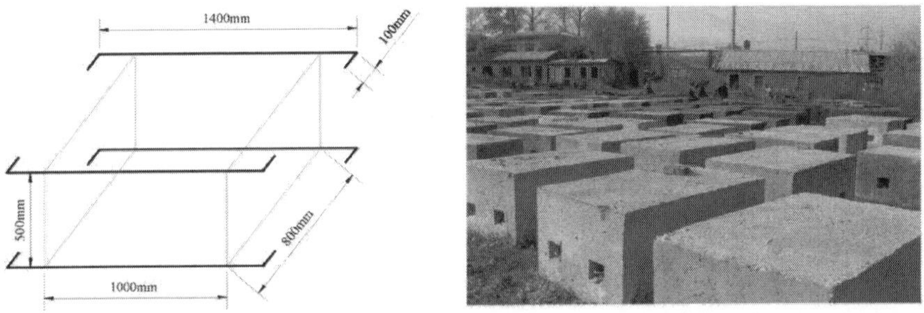

(a) (b)

Figure 17. Build-up of artificial side. (**a**) Anchor bar framework; (**b**) Large concrete blocks.

5.5. Performance of Artificial Side

5.5.1. Effect of GSE Retaining for the First Mining

Mining work started on 26 September 2012, at the Left No. 1 Working Face, and it finished on 25 October 2013. The length of the retaining roadway was 768 m. Deformation induced by the first mining work is shown in Figure 18.

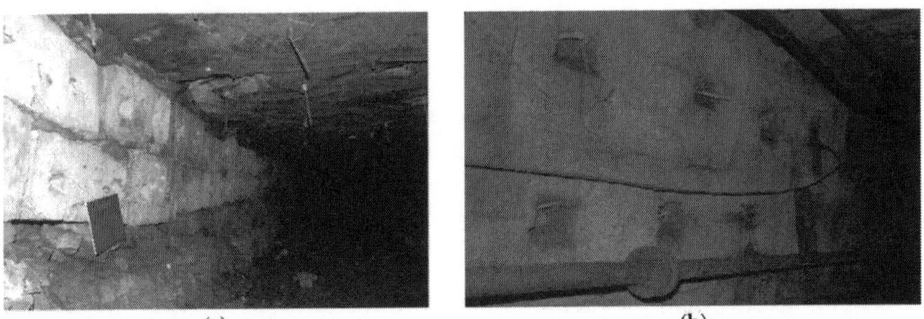

(a) (b)

Figure 18. Deformation due to the first mining. (**a**) Large deformation at 180 m behind working face; (**b**) Stabilization with shotcreting.

5.5.2. Effect of GSE Retaining for the Second Mining

The second mining work started on 17 March 2014, at the Left No. 2 Working Face, and was finished recently. Even after the second mining process, the AS was well-preserved intact condition except for partial cracks that appeared in the shotcreting layer. The AS withstood the second mining influence. Deformation induced by the second mining is presented in Figure 19.

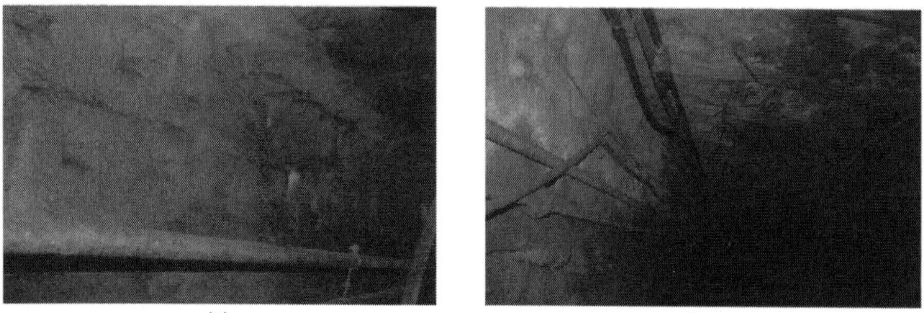

(a) (b)

Figure 19. Deformation due to the second mining. (**a**) Partial roadway; (**b**) Whole roadway.

5.5.3. Performance Evaluation

The AS was constructed and applied to the mining site to stabilize the entry. The AS with built-in anchor bar is still intact and has been stable for last three years, even after the two mining works. It can be seen that the anchor bar improved the load-carrying capacity of the AS and secured the stability of the working face.

6. CONCLUSIONS

An AS was introduced to stabilize the entry of a mining site and a failure criterion of AS was developed. A theoretical solution was derived to explain the failure mechanism, and a numerical study and experimental study were performed to validate the theoretical solution. An actual AS was installed at a mining site and its performance was validated.

An X-shaped failure pattern was found in the small and large-scale specimens. The X-shaped failure pattern obtained from experimental testing shows good agreement with the numerical simulation result. The failure pattern was clearly explained with a combination of two sets of principal shear stresses. Therefore, it can be concluded that the distribution of the maximum shear stress determines the failure pattern, regardless of specimen size.

Bolt-type reinforcement was introduced to enhance the load-carrying and anti-deformation capacity of artificial side. When the AS is stable, bolt-type reinforcement constrains the lateral deformation and improves the internal friction angle, cohesion *c*, and enhances the shear strength of the artificial side. When the AS is unstable due to damage, the constraining effect induced by bolt-type reinforcement can prevent a broken block from sliding along the shear failure plane and improve the residual strength of AS so that the load-carrying capacity can be sustained. In the field application, the reinforced AS was still stable even after mining operations for the last three years. It can be concluded that the anchor bar improved the load-carrying capacity of the AS and secured the stability of the working face.

ACKNOWLEDGMENTS

This research was supported by the National Basic Research Development Program (973 Program) of China (No. 2015CB251600), the National Natural Science Foundation of China (No. 51504182), the Natural Science Basic Research Plan of Shaanxi Province of China (No. 2015JQ5187), the Scientific Research Program funded by the Shaanxi Provincial Education Department (No. 15JK1466), the Project funded by China Postdoctoral Science Foundation (No. 2015M582685), and the Xi'an University of Science and Technology Research Fund Program for Young Scholars (No. 2014-NY-018). This research was also supported by INHA UNIVERSITY research grant.

AUTHOR CONTRIBUTIONS

Hong-sheng Wang and Dong-sheng Zhang conceived of the research. Hong-sheng Wang analyzed the data and wrote the paper. Lang Liu, Wei-bin Guo, Gang-wei Fan, KI-IL Song and Xu-feng Wang participated in the design of the study and verified the results. All authors have read and approved the final manuscript.

CONFLICTS OF INTEREST

The authors declare no conflict of interest.

REFERENCES

1. Wang, H. Research on the Creep Property and Its Stability Control Technology of Gob-side Entry's Narrow Side; China University of Mining and Technology: Xuzhou, China, 2011. (In Chinese)
2. Wang, H.; Zhang, D. Study on the Mechanisms and Application on Controlling the Creep Failure in Narrow Side of Gob-side Entry; China University of Mining and Technology Press: Xuzhou, China, 2015. (In Chinese)
3. Hu, X.; Yu, M. Research on triple-shear yield criterion for materials. Eng. Mech. 2006, 23, 6–11.
4. Zhang, L.; Gao, S.; Wang, Z.; Kong, L. Analysis of yield functions of marble under different loading and unloading stress paths. Chin. J. Rock Mech. Eng. 2014, 33, 2497–2503.
5. Niu, J.; Liu, G.; Tian, J.; Meng, L.-P. Comparison of yield strength theories with experimental results.Eng. Mech. 2014, 31, 181–187.
6. Yu, M. Twin shear stress yield criterion. Int. J. Mech. Sci. 1983, 25, 71–74.
7. Yu, M.; He, L.; Song, L. Twin shear stress strength theory and its promotion. Sci. China 1985, 28, 1113–1120.
8. Yu, M. Twin-Shear Theory and Its Applications; Science Press: Beijing, China, 1998. (In Chinese)
9. Yu, M.; Ma, G.; Li, J. Structural Plasticity; Zhejiang University Press: Hangzhou, China, 2009. (In Chinese)
10. Chen, L.; Sheng, G.; Chen, G. Investigation of impact dynamics of roof bolting with passive friction control. Int. J. Rock Mech. Min. Sci. 2014, 70, 559–568.
11. Chen, G.; Huang, L.; Chen, L.B.; Chang, J.-Y. Intelligent diagnosis of bolting systems under uncertain/dynamic impact. Microsyst. Technol. 2014, 20, 1733–1737.
12. Chen, G.S.; Chen, G.; Chen, L. Characterization of Dynamic Properties of Roof Bolts and their Applications. Adv. Mater. Res. 2014, 875–877, 599–604.
13. Wang, H.; Zhang, D.; Yang, J. Analysis on the dynamic effect of main roof's fracture and rotation and its application. Coalmine Saf. 2011, 42, 153–156.
14. China Academy of Building Research. JGJ 55-2011, Specification for Mix Proportion Design of Ordinary Concrete; China Architecture & Building Press: Beijing, China, 2011. (In Chinese)

15. Wang, H.; Zhang, D.; Fan, G. Structure effect of a soft-hard backfill wall in gob-side roadway. Min. Sci. Technol. 2011, 21, 313–318.

4

ALTERNATIVE PROCESS FLOW FOR UNDERGROUND MINING OPERATIONS: ANALYSIS OF CONCEPTUAL TRANSPORT METHODS USING DISCRETE EVENT SIMULATION

Jenny Greberg [1], Abubakary Salama [2], Anna Gustafson [1] and Bartłomiej Skawina [1]

[1]Department of Civil, Environment and Natural Resources Engineering, Division of Mining and Rock Engineering, Luleå University of Technology, Luleå SE-971 87, Sweden

[2]Department of Chemical and Mining, College of Engineering and Technology, The University of Dar es Salaam, Dar es Salaam P.O. Box 35901, Tanzania

ABSTRACT

As the near surface deposits are being mined out, underground mines will increasingly operate at greater depths. This will increase the challenges related to transporting materials from deeper levels to the surface. For many years, the ore and waste transportation from most deep underground mines has depended on some or all of the following: truck haulage, conveyor belts, shafts, rails, and ore pass systems. In sub-level caving, and where ore passes are used, trains operating on the main lower level transport the ore from ore passes to a crusher, for subsequent hoisting to the surface through the shaft system. In many mines, the use of the ore pass system has led to several problems related to the ore pass availability, causing production disturbances and incurred cost and time for ore pass rehabilitation. These production disturbances have an impact on the mining activities since they increase the operational costs, and lower the mine throughput. A continued dependency on rock mass transportation using ore passes will generate high capital costs for various supporting structures such as rail tracks, shaft extensions, and crushers for every new main level. This study was conducted at an existing underground mine and analyzed the transport of ore from loading areas at the lower levels up to the existing shaft points using trucks without employing ore passes. The results show that, when the costs of extending ore passes to lower levels become too great or ore passes cannot be used for production, haul trucks can be a feasible alternative method for transport of ore and waste up the ramp to the existing crusher located at the previous main level. The use of trucks will avoid installing infrastructure at the next main level and extending the ore passes to lower levels, hence reducing costs.

Keywords: rock mass transportation; haulage system; trucks; ore pass; discrete event simulation; sub-level caving

1. INTRODUCTION

As resources near the surface are being mined, underground mines worldwide are operating at increasing depths. The cost and time required for rock mass transportation increases with increased mining depths, and with the aim to increase mined volumes and reduce production costs, the rock mass transportation from deeper levels to the mine surface presents challenge for the underground mines. Despite the existence and availability of several haulage options, production in most deep underground mines depends on the safe and continuous operation of the ore pass system. Ore passes are used for material transportation, and can also serve as a means to store ore in underground mines [1]. However, in order to design a well-functioning ore pass, it is necessary to examine the entire ore handling system of the mine, from the production areas to the shaft points [2].

In sub-level caving, ore is mined on each sub-level from the hanging wall to the forefront of the footwall, starting with overlying sub-levels and proceeding downwards. As the ore is mined from a sub-level, the hanging wall collapses and covers the mining area with broken waste rock [3]. The ore is transported to the desired destination by either the use of ore passes, or by other hauling methods. When ore passes are used, trains operating on the main level transport the ore from the ore passes to the crusher, for subsequent hoisting to the surface through the shaft system. This incurs a high capital cost for various structures such as rail tracks, shaft extensions, and crushers for every new main level. Despite the existence of design guidelines, ore pass problems, including hang-ups and wall failures still persist in underground mines. These problems result in threats to the safety of personnel, loss of production, and an increase in the costs of repairs and production [4,5]. Even when the ore passes are carefully designed, the ore pass system is susceptible to several problems over its design life, including stability, wear, and material flow [6]. These problems increase delays in mining operations [7], and delays have an impact on the mining activities since they increase the operational costs and reduce the mine throughput. A study of more than 200 ore passes conducted in South African mines observed that more than 50 percent of these ore passes had stability problems, and 16 percent of them had been abandoned [8]. Several methods can be employed to restore the ore passes, such as the use of long-hole drilling and blasting, flushing a blocked ore pass with water, pushing rods, and the use of appropriate block size infrastructure to prevent the passage of oversized boulders [9]. In most cases, the actual cost of restoring ore passes is extremely high compared with the initial cost of development [8]. If ore pass restoration is not possible, and the ore pass therefore will not be used for production, and alternative transportation methods must be considered.

The aim of this study is to evaluate a conceptual future haulage method at an existing underground sub-level caving mine by comparing different haulage equipment of various sizes operating from the production drifts to the crushers, enabling rock mass transportation without using ore passes. Discrete event simulation was used for the analysis, and the study is based on data from an operating mine.

2. HAULAGE SYSTEMS FOR DEEPER LEVELS

Existing transport methods that are feasible for deep mines include truck haulage, shaft systems, and conveyor belts. Trucks are highly flexible in travel routes and fleet size, and provide high productivity [10]. Shaft systems may be inflexible because of their limited number of fixed feed points, but once installed they offer low operational costs [11,12,13]. Conveyor belts typically provide the most economic method for material transportation due to their high carrying capacity [13]. The choice of conveyor parameters is influenced by the nature of the material to be conveyed, the available tunnel space, and the overall economics of the system. Normally, all of these various haulage systems work in combination with other ore-handling components, such as Load-Haul-Dump (LHD) machines, ore passes and crushers. The haulage process is one of the most cost-intensive activities in a mining operation, and thus, one of the main contributors to operational costs [10,14]. The expected increase in transportation costs for deeper mines makes the choice of hauling equipment essential when seeking cost reductions for deeper level operations. The high cost of developing a new main level when a mine operates at great depth necessitates the technological innovation of an alternative haulage system. Frequent problems with ore pass failures, and the significant costs involved in managing these ore passes, make the continuation of the current ore handling approach questionable. While these challenges continue, mining operations seek to optimize the haulage system so that the desired production rates are achieved on time, and at minimum cost. This study evaluates the use of trucks to transport ore from the loading areas directly to the dumping point at the shaft without using ore passes. The use of trucks will help avoid new main level development and infrastructure costs, and hence reduce capital costs.

3. SIMULATION OF HAULAGE SYSTEMS

Underground haulage systems are complex, and involve high investment costs. For many years, analytical methods have been used to evaluate the performance of the system. These methods produce useful information on the system being analyzed, but their application is limited; for example, when the systems involve non-steady state conditions, or when operations involve uncertainties [14,15,16]. Under these conditions, an alternative method for analyzing a system is discrete event simulation. The random and dynamic nature of the haulage systems of mining operations makes them very difficult to model using analytical models. For that reason, in this work we have chosen to use discrete event simulation as the technique to evaluate the mine operations. When simulation is employed, model input can be based on appropriate probability distributions that characterize the input variables. The process of designing a model of a real system usually takes into account a set of assumptions for the operating system [17]. These assumptions are expressed in mathematical, logical, and symbolic relationships between the objects of the system of interest, and they can be solved analytically or by using simulation. With the aim of optimizing, improving, analyzing and planning of existing and future systems, various mining operations can be analyzed using simulation. These include: fleet requirements, the mine scheduling, and mine planning [14,17]. There are a number of simulation tools available in the market, such as SimMine, SLAM, SIMAN, ARENA, and AUTOMOD. In this study, the SimMine simulation software, which is based on discrete event simulation principles and has been developed specifically for the simulation of mining operations [18,19], is used.

4. CASE STUDY

4.1. Mine Description

This study was carried out in one of the largest underground mines in Sweden. The mine consists of a high-grade magnetite deposit approximately four kilometers long running in the northeasterly direction, with an average thickness of between 80 and 100 m. The mine uses the sub-level caving method (see Figure 1). In this method, development drifts are opened first, followed by the drilling of the ore passes. The ore passes extend vertically from the current mining area down to the bottom of a new mining area, where a transportation level is located. Horizontal sub-levels are created, including crosscuts that provide access to the production drifts. The self-supported crosscuts are drilled through the orebody, perpendicular to the access routes. The spaces between sub-levels are about 28.5 m, while spaces between crosscuts are 25 m. At the crosscuts, near-vertical rings of holes are drilled in a fan-shaped pattern. Each ring contains around 10,000 t of ore and waste. The ore is mined on each sub-level, starting with overlying sub-levels and proceeding downwards; in each sub-level, the ore is removed from the hanging wall to the forefront of the footwall. As the ore is mined from a sub-level, the hanging wall collapses by design, and covers the mining area with broken waste rock [3].

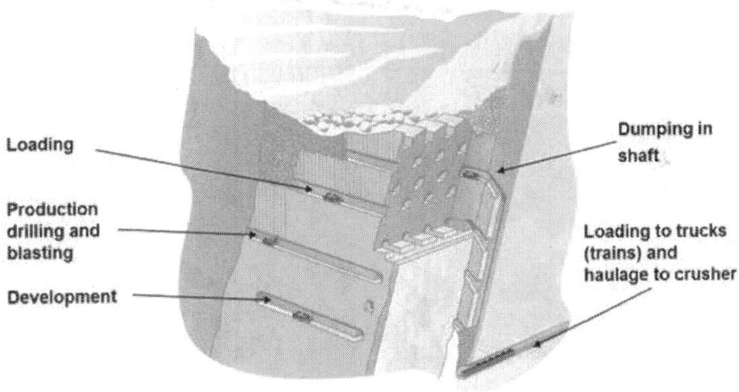

Figure 1. Sub-level caving mining method (Courtesy of Atlas Copco).

4.2. The Current Haulage System in the Mine

The mine is divided into 10 main production areas, called blocks, which extend from the uppermost mining level down to the current main level. Each mine block currently consists of 10 sub-levels. Each block is 400 to 500 m in length, and has its own group of ore passes located at the center of the production area and extending down to the main haulage level. Currently, the main haulage level is at 1365 m. Mining continues in each block using electric LHDs with a capacity of 25 t that operate from 6 a.m. until 10 p.m., while semi-automated diesel LHD machines with a capacity of 21 t operate from 10 p.m. to 6 a.m. Blasting normally occurs at around 00:00 every day. The LHDs load the ore from draw points within each production drift, and transport the ore to the ore passes. Large trains, operating on the main level, transport the ore from ore passes to a crusher. The crushed material is stored in the ore bins, and after that transported on a small conveyor belt to the hoisting system. To conform to mining restrictions, once mining begins in a block, it must be maintained until all available ore is removed before starting to mine the next one. The current loading operation from each block results in an average daily production of 6035 t. The future plan is to increase production to 37 Mt of crude ore per year from all ten blocks, which means a daily production of 10,000 t from each block.

4.3. Conceptual Haulage System

The next main level at the mine will be at the depth of 1685 m, followed by another level 2005 m deep. With increasing depth, stresses increase, creating higher risk for geomechanical problems such as ore pass failures. If ore pass restoration is not possible, and the ore passes cannot be used for production, alternative transportation methods have to be applied. In this study, a conceptual truck haulage system for transport of the material from the production areas to the crusher on the previous main level located at 1365 m was modeled and simulated. By using a truck haulage system, the installation of a rail-mounted track system can be avoided and the hoisting system will not have to be extended, thus reducing infrastructure costs.

4.4. Simulation Model

In order to identify the number and size of the trucks needed to reach the future production target, a simulation model was developed. The model formulation was done using the SimMine simulation software (SimMine AB, Malå, Sweden), which uses a full graphical user interface for model set up. It utilizes statistical distribution functions to model variations in process times. For verification purposes, and to increase understanding, the tool has a three-dimensional environment that offers animated visual feedback of the model, allowing viewing of the dynamic system as it operates [19,20].

4.5. Model Settings

The conceptual mine layout modeled in the simulation is shown in Figure 2. The modeled area consists of the current main level (1365 m) where the crusher is located, and a series of ramps connecting the production areas with the main level. In this study, two simulations were run, and can be described as follows:

1. Two sub-levels (located closest to the next main level at 1685 m) with two production areas in each level was considered. These two levels were selected for modeling since they are the levels located furthest away from the crusher on the previous main level, and hence require the largest number of trucks and the most time spent in queuing. These can be seen in Figure 3 as sub-levels 9 and 10.
2. Two other sub-levels (levels 14 and 15) situated below the main level at 1685 m were also simulated. This was done to analyze the effects of further increasing the distance to the existing crusher at the main level of 1365 m.

For the sub-levels above the main level at 1685 m, the modeled production areas (indicated as areas 1, 2, 3, and 4 in Figure 2) were chosen for the purpose of the truck modeling and simulation. Due to mining restrictions, this is the maximum number of production areas that can be used at the same time. Areas 1 and 3 are in the same production level, which for this study is called level 9, while areas 2 and 4 are on the level below, which is here called level 10. We also simulated four production areas (two at each level) for levels 14 and 15. The vertical distance between two production levels is 28.5 m.

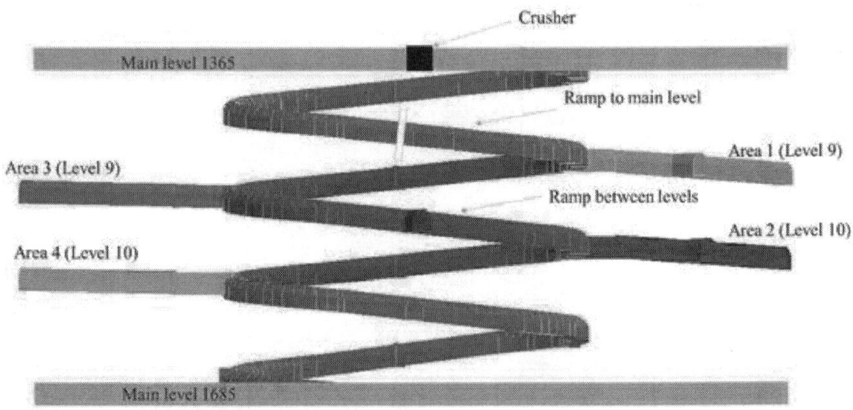

Figure 2. The conceptual mine layout.

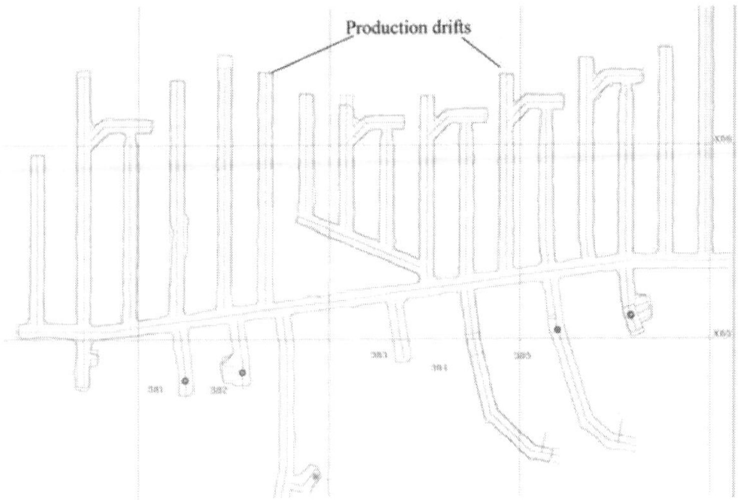

Figure 3. The production area considered in this study.

Each of the modeled production areas in the simulation consists of 17 productions drifts (see Figure 3), the total expected amount of ore to be mined being 6.17 Mt. According to the mine's regulations, production at level 10 starts when level 9 is 75 percent completed. This means that 1.54 Mt remains to be hauled from level 9, while 6.17 Mt will be hauled from level 10. There is no sub-level below level 10 to commence production when the lower level is 75 percent complete, so the results that originate from the last 25 percent of level 10 will show less traffic disturbance. The reason is that the next level below it is the new main level at 1685 m.

Haulage is done by trucks from all production areas to a crusher at the main level (1365 m). Trucks are loaded at the loading chamber and use a ramp to transport material to the crusher. The simulation model also includes the time loss when trucks meet to give way to each other, when they meet at intersection points between ramp and crosscuts, and when they meet at the corner points. The model logic applies the rule that when trucks meet, the empty truck gives way to the loaded truck at the nearest waiting zone. The haulage distances from each ramp to the entry point at the main level (1365 m) are 2320, 2610, 4060 and 4350 m for levels 9, 10, 14 and 15 respectively. The length of the main level is 2150 m, measured from the entry of the ramp to the crusher. This makes the total haulage distances traveled by trucks 4470, 4760, 6210 and 6500 m for levels 9, 10, 14 and 15 respectively. The crusher can be fed from two dumping points. During simulation, it was assumed that the crusher would never break down. The simulation was run until the machines in operation finished removing all the available ore from the simulated production areas. All ramps have an inclination of 1:10.

4.6. Scenarios

Seven scenarios were evaluated and analyzed, all aiming at meeting the stated production target of 10,000 t per day per production area (Table 1). Different combinations of machines were used for each scenario. Two sizes of LHDs with capacities of 10 and 21 t and five different sizes of trucks, with capacities of 20, 21, 40, 42 and 63 t were used.

Table 1. Scenarios.

Scenario	LHD (Load-Haul-Dump) Bucket Capacity (Tonnes)	Truck Capacity (Tonnes)	Sub-Levels
1	21	21	9 and 10
2	21	42	9 and 10
3	21	63	9 and 10
4	21	21	14 and 15
5	21	42	14 and 15
6	10	20	9 and 10
7	10	40	9 and 10

For all scenarios, both electric and diesel trucks were used, the number of electric LHDs was limited to two at each production area. When two LHDs were employed, each one of them worked on a different side of the production area, meaning that there was no interaction between the LHD machines during production. The LHDs were used to load the trucks at each working area. The rationales behind the selection of the different scenarios are as follows:

- Scenario 1 was selected to analyze the haulage system when two LHDs serve smaller sized trucks (with capacities of 21 t).
- Scenarios 2 and 3 were selected to analyze the effect of traffic and queuing on loading and dumping points for each production area when bigger units (trucks with capacities of 42 and 63 t) are employed.
- Scenarios 4 and 5 were selected to simulate the effects of increases in depth when having to transport the ore from deeper levels (14 and 15) to the existing dump station at the current main level at 1365 m.
- Scenarios 6 and 7 were selected to allow for smaller drift sizes, such that larger trucks cannot fit. Therefore, two electric LHDs, each with a bucket capacity of 10 t, were used to load trucks with box capacities of 20 and 40 t.

In all scenarios, the aim was to determine the number of trucks needed to reach the desired production goals. In scenarios 1 to 5, operations with large drifts/openings (5 m wide) were modeled and analyzed, whereas in scenarios 6 and 7, operations with small drifts/openings (less than 5 m wide) were modeled and analyzed. Note that in scenarios 1 and 4, one bucket is assumed to fill the truck. This would in the real operation create a risk of overloading the truck. Still, these scenarios are presented here since they are used to illustrate what happens when a smaller truck is used.

4.7. Input Data

The cases were analyzed considering variations in the availability of trucks and production areas, and also disturbance from other mining vehicles on their way to and from the production haul areas. The truck availability was assumed to be 90 percent. The production area availability was set at 100 percent for areas with fewer interactions, and 80 percent for areas with several other mining activities. The availability of LHD machines was set at 100 percent for two of the cases, since the LHD was immediately replaced after breakdown with another during the course of production, and 90 percent when down time was considered. Keeping a back-up unit can be a costly strategy especially for a small mining operation. However, for a large mining operation like the one presented in this paper, this is a feasible strategy and therefore used in this study.

According to the data obtained from the mine, the disturbance from other mine vehicles to the trucks was modeled as somewhere between zero percent and two percent. This means that up to two percent of the total available time is lost due to traffic disturbance, and hence the delay for trucks increases when the mine vehicles are driving up or down the ramp and along the main level. The term 'traffic disturbance' refers to the percentage of time assumed to be lost when trucks meet in the haul ways (the main drift and the ramp). No traffic disturbance occurs inside production areas because within the production drifts, only LHDs are travelling to transport material from production faces to a hauling point. Input data for the simulation were collected from the mine, and from the equipment manufacturers. Five different cases were studied for each scenario (Table 2).

Table 2. Input parameters for each case.

Cases	LHD Availability (%)	Truck Availability (%)	Area Availability (%)	Traffic Disturbance (%)
1	100	90	100	0
2	90	90	100	0
3	100	90	80	0
4	90	90	80	0
5	90	90	80	2

The assumed truck parameters (Table 3) used in the simulation are based on data from the mine and from manufacturers. The data for the 21 t truck were obtained from a different mine since the 21 t truck was not used in the studied mine. There is a great variability in the loading time of the hauling units. To determine the proper distribution function, a statistical analysis was performed and a triangular distribution was selected to model the loading times. The size of the production drifts does not accommodate the truck size and therefore, during loading, the trucks wait at the access point that connects the main level and the production drifts. The loading time includes the time taken by the LHDs to travel from the production drifts to the truck loading point. The loaded truck moves up the ramp, which has an incline of 1:10, while the empty truck moves down the ramp decline to the loading points.

Table 3. Truck parameters.

Machine	Truck Capacity (t)	Loading Time (s)	Dumping Time (s)	Speed when Empty (km/h)	Average Speed when Loaded (km/h)
Volvo 21 t Truck	21	Tri (10,12.5,15)	10	30	18 incline 30 horizontal and decline
Atlas Copco MT42	42	Tri (145,162,179)	10	30	11.3 incline 30 horizontal and decline
Atlas Copco MT6020	63	Tri (450,505,567)	10	30	12 incline 30 horizontal and decline
Volvo 20 t Truck	20	Tri (143,167,180)	10	30	18 incline 30 horizontal and decline
Atlas Copco MT42	40	Tri (403,452,507)	10	30	11.3 incline 30 horizontal and decline

Tri = triangular distribution Tri (a,m,b) where m is the most likely value.

4.8. Verification and Validation

Simulation model verification is the process of ensuring that the model design has been transformed with sufficient accuracy into a computer model [21]. Validation is the process of ensuring that the model is sufficiently accurate for the purpose intended [21]. The stages for verification and validation processes used in this study are based on the one proposed by Robinson [22]. As seen in Figure 4, the stages included in the process are:

- Conceptual model validation. This stage determines if the scope and level of detail of the proposed model are sufficient for the purpose intended, and that the assumptions are correct.
- Data validation. This stage determines that the data used for building the model, for validation and for experimentation, are accurate.
- White box validation. This stage determines that the components of the computer model accurately represent the corresponding real world elements.
- Black box validation. This stage determines that the overall model represents the real world with sufficient accuracy.

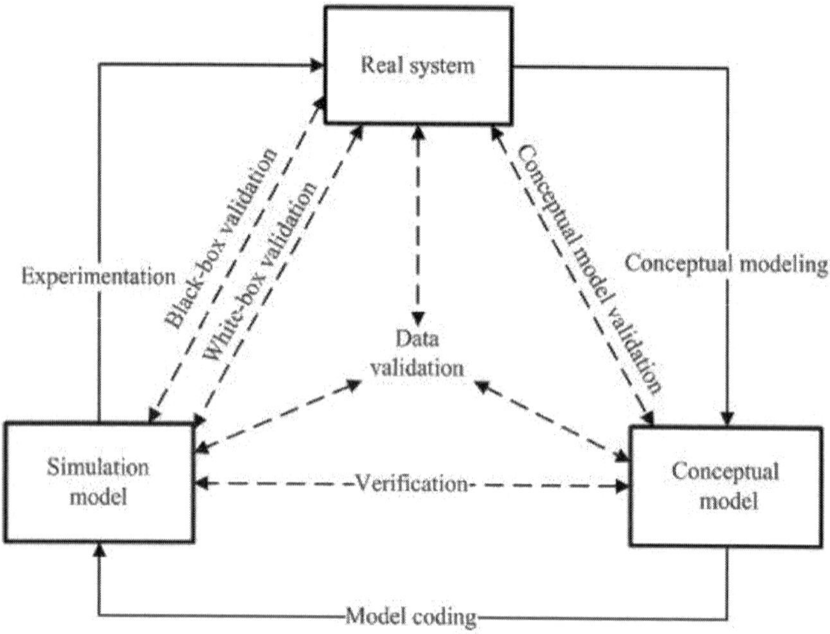

Figure 4. Verification and validation in the modeling process (modified from source [22]).

The model is considered "valid" when the assumptions underlying the conceptual model are correct, and when it has been determined that the model represents the real system [15,21,23].

The data validation was performed according to the steps described below:

1. Production data from the operating mine together with machine specifications were used to derive and calculate output parameters of interest. The data used were actual t/h, shift schedule, equipment availability. The calculations resulted in specific values for number of working hours required to load the ore from a production area, number of required days required to load the ore from a production area and number of hours the equipment were unavailable.
2. The next step was to run a simulation using the same input data and boundary conditions as for the calculations in step 1. The simulation model uses more detailed equipment data (speed, time to load, time to dump, as presented in Table 3) based on the mine operating data and data from the manufacturer.
3. The results from the simulation were then compared with the results from the calculation in step 1, ensuring that the simulation model gave similar output results.
4. A mathematical analysis was also done when calculating and verifying the theoretical cycle time for trucks (drive to the dump location, dump, and drive back again), to verify the upper limit of the trucks.

Black box validation was then carried out by comparing the output from the simulation model with the output from the real system. This was done by comparing the actual hourly production of the mine with the simulation output for one mine block (shown in Figure 3). The average actual hourly production for this block is 4147 t/h with the 21 t loader. The output from the simulation model resulted in an hourly production of 41,474 t/h with the 21 t loader, which validated the model and proved that it behaved like the real system.

White box validation was carried out by using debugging techniques, animations, model inspections by specialists, and by running the model under varying conditions.

5. RESULTS AND DISCUSSION

The simulation was run based on mining in the 9th, 10th, 14th, and 15th levels in the mine (see Figure 2). At each level, trucks were used to transport the material from loading areas to the crusher, which is located at the current main level at 1365 m. Each level consists of two production areas. Two electric LHDs, with bucket capacities of 21 and 10 t, were used to load the trucks at each production area. The 21 t LHDs were used to provide in the case of the large openings, while 10 t LHDs were used for smaller openings. The aim of the simulation was to determine the number of trucks needed to meet the production target of 10,000 t when the ore pass system is not used.

5.1. Trucks Needed

The simulation was first conducted for the case of 21 t LHDs loading 21 t trucks (scenario 1). After the first simulation, the model was adjusted, and the other scenarios shown in Table 1 were simulated. The results for scenario 1 show that, when using smaller sized LHDs to load onto smaller sized trucks, between 13 and 16 trucks are needed, depending on the case being analyzed (see Figure 5). In case one, fewer trucks (13) are required to meet the target compared with the other cases. This is because, for this case, we assumed that LHDs were immediately replaced with another after breakdown during production, and we also assumed that there could be no disturbances from other mine operations in the production area or with other mine vehicles on the ramp.

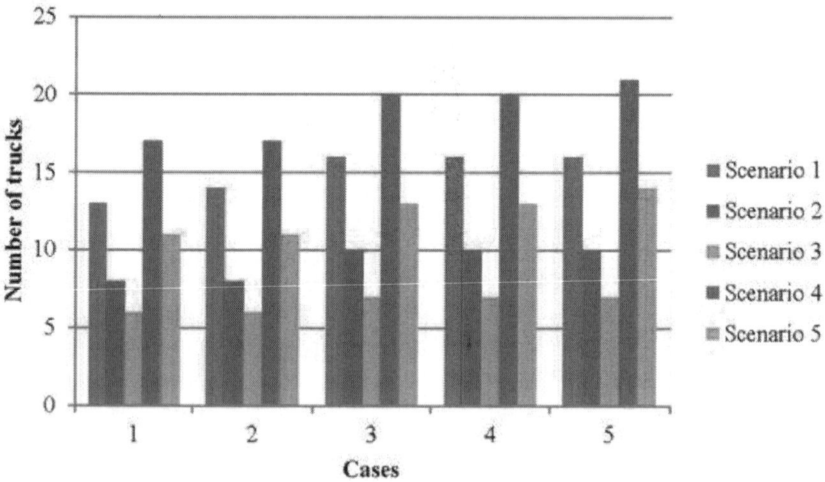

Figure 5. Truck requirements for large openings.

The results for scenarios 2 and 3 show that, when larger trucks are used, fewer machines (between eight and ten for scenario 2, and between six and seven for scenario 3) are required to achieve the production goal. For these scenarios, the truck sizes used were 42 and 63 t, respectively. The reason for requiring fewer trucks is that, although the bigger trucks have a slightly slower speed, over the same time period they dump more material than smaller sized trucks.

The results for scenarios 4 and 5 show that more trucks are needed when mining deeper: between 17 and 21 trucks for scenario 4, and between 11 and 14 trucks for scenario 5. The reason for this is that, under these scenarios, trucks have longer cycle times since they have a longer distance to travel to the dump station (6210 m for level 14 and 6500 m for level 15). Compared with scenarios 1 and 2, more trucks were required to achieve the daily production under scenarios 4 and 5, despite the fact that the compared scenarios use the same truck sizes. Cases 3 and 4 result in similar number of trucks for all scenarios despite the difference in truck and loader sizes. This is because for these cases, area availability was reduced to 80%, which leads to more idle time for loading and haul units.

The results for scenarios 6 and 7 are shown in Figure 6. In these scenarios, electric LHDs with a capacity of 10 t were used to load the trucks with capacities of 20 and 40 t. It shows that a higher number of 20 t trucks were needed (between 17 and 24) compared with 40 t trucks (between 10 and 16).

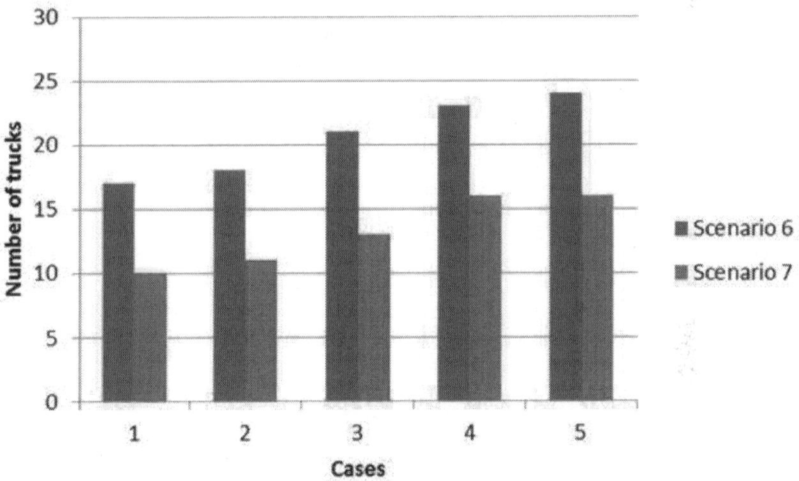

Figure 6. Truck requirements for small openings.

For example, in scenario 6, case 5, more than 20 trucks were required to attain the production target due to presence of higher traffic disturbance when a large number of trucks are in operation. For scenario 7, although case 5 also allows for two percent of time loss due to traffic, fewer trucks were needed to meet the target. The reason for this is that, when production goes ahead in this case, no trucks are available for production in other areas. This, therefore, reduces the congestion of trucks on the ramp.

Comparing all the scenarios, the one requiring the fewest trucks to achieve the daily production target was scenario 3. This is because trucks with the highest capacity were used for this scenario. For cases 3, 4 and 5, since the area availability is lower, a large number of trucks is needed to meet the daily production target. Furthermore, based on the results from scenarios 4 and 5, it can be shown that, the deeper the mine gets, the more trucks are needed.

5.2. Truck Traffic on the Ramps

During simulation, the different haulage options were evaluated by determining the truck traffic at each production area depending on which level the trucks were working. The term traffic refers to the percentage of time lost when trucks meet in the haul ways (the main drift and the ramp). When trucks meet at the intersection points between ramp and crosscuts, or at the corner points, the empty truck gives way to the loaded truck at the nearest wait zone. As seen in Figure 7 for levels 9 and 10 (scenarios 1, 2, 3, 6 and 7), the average time loss due to queuing and waiting for trucks on the ramps is 8.17 percent. Scenario 6 generates higher truck traffic on the ramps due to the depth of levels 9 and 10, and also due to the large number of trucks in operation compared with other scenarios. The third scenario generates less traffic on the ramps since it requires the fewest trucks to meet the required production target. The average transportation time rose to over 23 percent of the total working time when the smaller sized LHDs were used to load the trucks (levels 9 and 10).

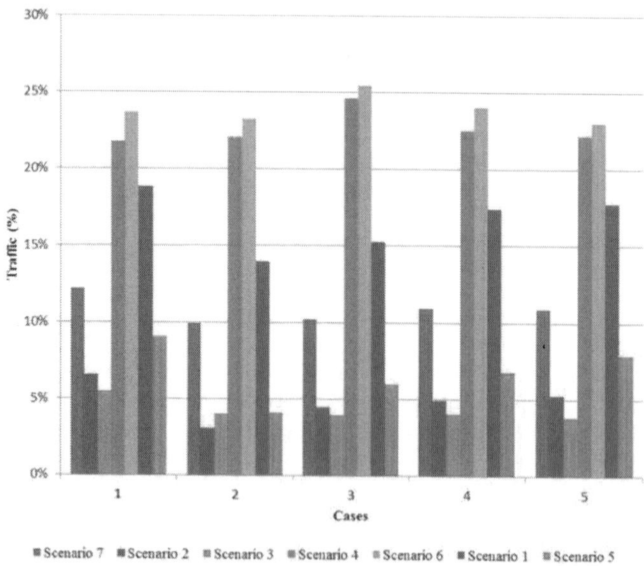

Figure 7. Truck traffic for all scenarios.

Since more trucks are needed at levels 14 and 15 due to increased depth, the truck traffic on the ramps increases, and hence results in increased idle time for each truck due to queuing problems. This shows that, when deeper levels are in operation, a lot of time is lost due to queuing and waiting time for trucks on the ramps.

An analysis of the results indicates that the most favorable option for large openings is the haulage system using trucks with a capacity of 63 t. The results also show that the favored option for smaller openings is to use trucks with a capacity of 20 t. To achieve the planned production target of 10,000 t per day from each mine block, the option with large openings requires about seven trucks to be in operation, whereas more than 20 trucks would be required for smaller openings. This shows that, for smaller production drifts, hauling using trucks will not be an economical option in order to attain the target of 10,000 t per day.

Diesel truck haulage systems are highly flexible in terms of travel routes and fleet size, and present no electrical hazards. However, the use of a large number of trucks underground increases the risks from flammable fuel, results in higher heat emissions and noise levels, and emits toxic gases into the mine environment. This results in higher costs because of the additional energy needed for longer haul distances, and the need for additional ventilation to mitigate geothermal heat and exhaust gas emissions.

Based on the ratio of emissions and the amount of fuel used, the diesel machines emit 2.68 kg of CO_2 gas for every liter of diesel fuel used [24]. The need for ventilation to mitigate engine heat and emissions will be higher, and this will lead to increased operating costs. One of the possible alternatives when using smaller trucks is to develop two ramps: one ramp will be used by loaded trucks (incline) and the other by empty trucks (decline).

For large openings, the analysis shows that seven trucks can be employed to attain the planned production target. More than seven trucks will increase traffic congestion, result in longer idle times, higher operating costs, and lower the production rate.

Due to increased depth, and hence increased rock stresses, problems with ore passes failure are expected, which will lead to further delays in material transportation. In sub-level caving, when ore passes and trains are used to transport the ore to a crusher, ore pass management is necessary in order to avoid interruptions to material flow. When the costs of managing the ore passes become too high, or when ore passes cannot be used for production, haul trucks can be employed as an alternative method to transport ore and waste up the ramps to the existing crusher located at the previous main level. If truck haulage is adopted, a crusher and a rail-mounted track system at the lower main level will no longer be required, and there will be no need to extend a shaft from the previous levels. This will reduce the reinvestment costs for the new main level infrastructures.

5.3. General Discussion of the Results and Method

The choice of using simulation to solve problems is sometimes disputable and in many cases analysis using simulation is too complex and time consuming given the nature of the study. For this specific study, using simulation was the preferred option, and the results presented above could not have been generated with the same accuracy through manual calculations alone. Analyses of the given problem include more input information than could be handled by manual calculation and still generate results with the same degree of detail and accuracy as possible with a simulator. In the simulations concerning level 9 and 10, the output results also showed important information on how the traffic disturbances affected the overall performance and the variance in total time each truck was working in one area until it started to drive on the ramp (the trucks driving time and loading time) which affects the frequency at which trucks entered the ramp. The number of trucks also affected queuing, both inside each area and at the dump location, and the availability of the loaders is another important factor that affects how the trucks work.

All of these factors, which are hard to describe and calculate since they are not deterministic values or constants, require a dynamic model that behaves like reality to obtain a reliable result. For the purpose of investigating a possible future scenario, the most appropriate tool possible should be used, which in this case is a simulator.

6. CONCLUSIONS

This study was conducted in an existing deep underground mine that uses sub-level caving. Discrete event simulation was used to evaluate options for future haulage from the lower mine levels without the use of ore passes by comparing different haulage equipment with various sizes operating from the production drifts to the crushers. The results indicated the following:

- Haul trucks can be employed as an alternative haulage method when the costs of managing the ore passes become too high, or the ore passes cannot be used for production.
- For large openings, the favored option for this case is a haulage system using trucks with a capacity of 63 t.

- For smaller production drifts, hauling using trucks will not be economical in order to attain the desired production target.

The study showed that discrete event simulation together with verification and validation processes to develop a model for analysis is a suitable tool for investigating and analyzing mine operations prior to new investments or implementation of new systems. There are, however, many additional considerations that analysis by simulation may not be able to resolve. To evaluate additional costs and considerations when using trucks, discrete event simulation can be combined with economic analysis models to improve understanding and reduce risk related to the selection of the operational systems.

Future research in this area will focus on combining discrete event simulation and mixed integer programming to study the following: ore pass management costs, additional truck operating costs, new haulage systems for deeper levels, and the capital costs required for installation of the necessary infrastructures in the new main level. Future work on comparison of the capital and operational cost of electric versus diesel trucks is also suggested.

ACKNOWLEDGMENTS

This study is part of the I2 Mine Project (Innovative Technologies and Concepts for the Intelligent Deep Mine of the Future), WP 2 subtask 2.1.1, and has been carried out at the Division of Mining and Geotechnical Engineering at Luleå University of Technology. The project is funded by EU 7th framework programme.

AUTHOR CONTRIBUTIONS

Jenny Greberg and Abubakary Salama conceived and designed the experiments/simulation. Abubakary Salama, Jenny Greberg and Bartlomiej Skawina performed the experiments by simulation. Jenny Greberg, Abubakary Salama, Bartlomiej Skawina and Anna Gustafson analyzed the data. Jenny Greberg and Abubakary Salama were the main authors of the paper, assisted by Anna Gustafson and Bartlomiej Skawina.

CONFLICTS OF INTEREST

The authors declare no conflict of interest.

REFERENCES

1. Stacey, T.R.; Wesseloo, J.; Bell, G. Predicting the stability of rock passes from the geological structure.J. S. Afr. Inst. Min. Metall. 2005, 105, 803–808.
2. Stacey, T.R.; Swart, A.H. Investigation into Draw Points, Tips Orepasses and Chutes; Technical Report Report to the Safety in Mines Research Advisory Committee, Project OTH 303; Mine Health and Safety Council: Johannesburg, South Africa, 1997; Volume 1.
3. Kuchta, M.; Newman, A.; Martinez, M. Long- and short-term production scheduling at LKAB's Kiruna mine. In Handbook of Operations Research in Natural Resources; Weintraub, A., Romero, C., Bjørndal, T., Epstein, R., Eds.; Springer: New York, NY, USA, 2007; pp. 579–593.
4. Iverson, S.R.; Jung, S.J.; Biswas, K. Comparison of orepass computer simulations for design against dynamic load. In Proceedings of the SME Annual meeting, Cincinnati, OH, USA, 26–28 February 2003.
5. Brummer, R. Design of Ore Passes Methods for Determining the Useful Life of Ore-Passes Based on Previous Experience and Case Studies; Technical Report CAMIRO; Mining Division Limited: Sudbury, ON, Canada, 1998.
6. Hadjigeorgiou, J.; Esmaieli, K.; Harrisson, R. Observation of ore pass system performance at Brunswick Mine. CIM Bull. 2008, 101, 1–13.
7. Hadjigeorgiou, J.; Stacey, T.R. The absence of strategy in ore pass planning, design and management.J. S. Afr. Inst. Min. Metall. 2013, 133, 795–801.
8. Joughin, W.C.; Stacey, T.R. Risks associated with rock passes in deep-level tabular mines based on historical pass performance. J. S. Afr. Inst. Min. Metall. 2005, 105, 795–802.
9. Hadjigeorgiou, J.; Lessard, J.F. Strategies for restoring material flow in ore and waste pass systems.Int. J. Min. Reclam. Environ. 2010, 24, 267–282.
10. Kecojevic, V.; Komljenovic, D. Haul truck fuel consumption and CO2 emission under various engine load conditions. Miner. Eng. 2010, 62, 44–48.
11. Atkinson, T. Selection and sizing of excavating equipment. In SME Mining Engineering Handbook, 2nd ed.; Hartman, H.L., Ed.; SME: Littleton, CO, USA, 1992; pp. 1311–1333.

12. McCarthy, P.L. Selection of shaft hoisting or decline trucking for underground mines. In Driving Down Haulage Costs; AMC Reference Library: Kalgoorlie, Australia, 1999.

13. Pratt, A.G.L.; Ellen, P.J. Selection of an ore haulage system for Telfer deeps. In Proceedings of the Hoist and Haul Conference, Melbourne, Australia, 5–7 September 2005; Australasian Institute of Mining and Metallurgy: Perth, Australia, 2005; pp. 131–140.

14. Salama, A.; Greberg, J.; Schunnesson, H. The use of discrete event simulation for underground haulage mining equipment selection. Int. J. Min. Miner. Eng. 2014, 5, 256–271.

15. Banks, J.; Carson, J.S.; Nelson, B.L.; Nicol, D.M. Discrete Event System Simulation; Pearson Education: Upper Saddle River, NJ, USA, 2010.

16. Raj, M.G.; Vardhan, H.; Rao, Y.V. Production optimization using simulation models in mines: A critical review. Int. J. Oper. Res. 2009, 6, 330–359.

17. Sturgul, J.R. Discrete mine system simulation in the United States. Int. J. Min. Reclam. Environ. 1999,13, 37–41.

18. Salama, A. Haulage Systems Optimization for Underground Mines: A Discrete Event Simulation and Mixed Integer Programming Approach. Ph.D. Thesis, Luleå University of Technology, Luleå, Sweden, 2014. pp. 1402–1544.

19. Greberg, J.; Sundqvist, F. Simulation as a tool for mine planning. In Proceedings of the 2nd International Future Mining, Sydney, Australia, 22–23 November 2011; pp. 273–278.

20. SimMine, Version 1.19. Available online: http://simmine.com/ (accessed on 29 June 2016).

21. Sargent, R.G. Verification and validation of simulation models. In Proceedings of the 37th Winter Simulation, Syracuse, NY, USA, 4–7 December 2005; pp. 130–143.

22. Robinson, S. Simulation model verification and validation: Increasing the user's confidence. In Proceedings of the 29th Winter Simulation Conference, Atlanta, GA, USA, 7–10 December 1997; IEEE Computer Society: Washington, DC, USA, 1997.

23. Balci, O. Validation, verification, and testing techniques throughout life cycle of a simulation study.Ann. Oper. Res. 1994, 53, 121–173.

24. Salama, A.; Greberg, J.; Skawina, B.; Gustafson, A. Analysis of energy consumption and gas emissions for loading equipment in underground mining. CIM J. 2015, 6, 179–188.

5

SPATIO-TEMPORAL MODELLING OF DUST TRANSPORT OVER SURFACE MINING AREAS AND NEIGHBOURING RESIDENTIAL ZONES

Lubos Matejicek [1,*], Zbynek Janour [2], Ludek Benes [3], Tomas Bodnar [3] and Eva Gulikova [4]

[1]Institute for Environmental Studies, Faculty of Science, Charles University in Prague, Benatska 2, Prague 2, CZ 12801, Czech Republic

[2]Institute of Thermomechanics, Academy of Sciences, Dolejskova 5, Prague 8, CZ 18200, Czech Republic

[3]Department of Technical Mathematics, Czech Technical University in Prague, Karlovo Namesti 13, CZ 12135, Czech Republic

[4]Ecoprogress, Zatecka 1899, Most, CZ 43401, Czech Republic

ABSTRACT

Projects focusing on spatio-temporal modelling of the living environment need to manage a wide range of terrain measurements, existing spatial data, time series, results of spatial analysis and inputs/outputs from numerical simulations. Thus, GISs are often used to manage data from remote sensors, to provide advanced spatial analysis and to integrate numerical models. In order to demonstrate the integration of spatial data, time series and methods in the framework of the GIS, we present a case study focused on the modelling of dust transport over a surface coal mining area, exploring spatial data from 3D laser scanners, GPS measurements, aerial images, time series of meteorological observations, inputs/outputs form numerical models and existing geographic resources. To achieve this, digital terrain models, layers including GPS thematic mapping, and scenes with simulation of wind flows are created to visualize and interpret coal dust transport over the mine area and a neighbouring residential zone. A temporary coal storage and sorting site, located near the residential zone, is one of the dominant sources of emissions. Using numerical simulations, the possible effects of wind flows are observed over the surface, modified by natural objects and man-made obstacles. The coal dust drifts with the wind in the direction of the residential zone and is partially deposited in this area. The simultaneous display of the digital map layers together with the location of the dominant emission source, wind flows and protected areas enables a risk assessment of the dust deposition in the area of interest to be performed. In order to obtain a more accurate simulation of wind flows over the temporary storage and sorting site, 3D laser scanning and GPS thematic mapping are used to create a more detailed digital terrain model. Thus, visualization of wind flows over the area of interest combined with 3D map layers enables the exploration of the processes of coal dust deposition at a local scale. In general, this project could be used as a template for dust-transport modelling which couples spatial data focused on the construction of digital terrain models and thematic mapping with data generated by numerical simulations based on Reynolds averaged Navier-Stokes equations.

Keywords: 3D laser scanning; GPS; Spatio-temporal modelling; Dust transport; GIS; Spatial database; Visualization; RANS equations

1. INTRODUCTION AND THE GENERAL APPROACH

The impacts of mineral dust on the natural environment in the neighbourhood of surface coal mines have long been recognized. One prerequisite for estimating these impacts is a determination of wind flows over the precisely mapped surface. Despite existing studies on the subject [1-3], practical numerical calculations have not been possible until recently. This is because the processes governing emissions and transport by wind flows over a surface are very complex. There is a dependency on terrain morphology which causes difficulties in estimating the dust emission rate and determining dust transport. In addition to wind flows, a set of physical processes in soils predetermines the conditions of primary dust dispersion. These processes are affected by highly variable natural factors, such as climate, soil state, and surface roughness. In cases of coal mine surfaces, other factors include mining activities, especially temporary storage, coal sorting, excavators, conveyors, and moving vehicles. The transport of coal dust is characterized by turbulent interaction in the atmospheric boundary layers. The dust particles are often driven by intensive meso-scale to synoptic-scale systems over long distances. At the more local spatial scale of the surface coal mine and neighbouring areas, a more precise spatio-temporal modelling of the mine surface and wind flows are needed for better predictions of coal dust transport.

In order to achieve this spatio-temporal modelling of dust transport over the surface of a mining area and neighbouring residential zones, a geographic information system (GIS) is used to create spatial data structures and modelling tools. Then, numerical simulations of the wind flows supported by a digital terrain model (DTM) can be carried out more precisely. In order to carry out this advanced spatial analysis, recent GIS are used to manage complex data structures and to create spatial models. Finally, visualization methods assist in the creating of map compositions for the presentation of modelling results. Running of GIS on advanced computer systems allows the implementation of these types of time consuming numerical algorithms focused on environmental research [4, 5]. The GIS development tools can assist in creating domain-specific components dedicated to data exchange or numerical modelling.

This paper describes GIS methods for the construction of a DTM and visualization of dust dispersion based on numerical modelling. In order to provide spatio-temporal modelling, a GIS database is created to manage data from 3D surface laser scanners, GPS measurements, existing thematic maps, numerical models and meteorological stations. The system is then applied to a simulation of dust transport over part of a surface coal mine located in the Czech Republic.

2. THE MAIN DATA SOURCES FOR THE SPATIO-TEMPORAL MODELLING

Spatio-temporal modelling is generally considered as a synthesis of GISs, which describe the spatial environment, and environmental models, which are linked to dynamic processes in the natural environment. Environmental models require data about the environment within which the processes occur, and simulations of environmental models provide additional data to extend the environmental descriptions. Thus, GIS can be used to serve as a common analysis framework for environmental models. Originally, GIS and environmental modelling were developed separately, which is indicated by their different data structures and functionality for data exchange and the user interface. Nowadays, however, there are some strategies for how to integrate environmental modelling in the GIS environment. They range from simple data exchange through shared data files to developing analytical tools into fully functional GISs [6]. In the case of spatio-temporal modelling of dust transport, the whole range of methods and functions can be solved using existing GIS capabilities, but extensions are necessary to process the numeric simulations of wind flows.

The main spatial data input for wind flow simulation is a DTM that is based on 3D surface laser scanning, GPS measurements and existing thematic mapping. The specific selection of spatial data inputs for the DTM and visualization was carried out depending on the required precision and actual availability.

2.1. Data sources from surface laser scanning

3D laser scanners provided an efficient method for local 3D point cloud acquisition of the surface coal mine and its associated industrial installations. Complete coverage with point cloud data mostly requires data acquisition from multiple standpoints. In order to improve existing thematic spatial data, four scans with an average grid size of 0.2 meters were performed with a Leica HDS 3000 3D laser scanner. The selected local area of the surface coal mine is shown in Figure 1. The temporary storage site located on the left represents the dominant dust emission source. The residential zone located approximately 500 meters from the temporary storage place is on the right side. The measurement equipment at one of the standpoints is illustrated in Figure 2.

Figure 1. The local area for 3D surface laser scanning (based on the aerial images: Mapy.cz).

In order to create the required spatial data inputs for such a complex DTM construction [7], processing of the point clouds was performed using Leica Geosystems HDS Cyclone software in cooperation with Gefos Ltd. in the Czech Republic. The final 3D point clouds for the area are shown in Figure 3, and a detailed view of the temporary storage site is in Figure 4.

Figure 2. The 3D laser scanner Leica HDS3000 making point cloud acquisition for a slope area on the left, and for the temporary storage place on the right.

Figure 3. The final 3D point clouds for the local area processed by Leica Cyclone software.

Figure 4. A more detailed view of the temporary storage and sorting place with the coal mine installations and other equipment.

2.2 Data sources from GPS

While the 3D surface laser scanning was used for mapping of continuous parts, a GPS Trimble-Pathfinder ProXT was used to assist in the capturing of the lines of the surface mine slopes, the boundaries of the temporary storage site and the locations of other surface objects (transport routes, other potential emission sources and meteorological stations). The basic measurement scheme is shown in Figure 5. The accuracy of the GPS measurements was improved in the postprocessing phase using data from the nearest reference station in the Czech Republic (Karlovy Vary, Figure 6). The final accuracy of the spatial data after the postprocessing phase was 0.5 m.

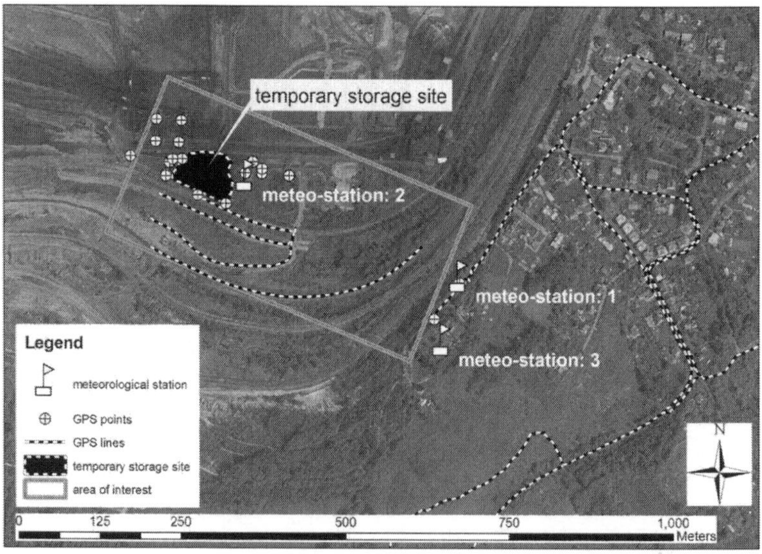

Figure 5. The lines of the surface mine slopes, the boundaries of the temporary storage site and the locations of other surface objects.

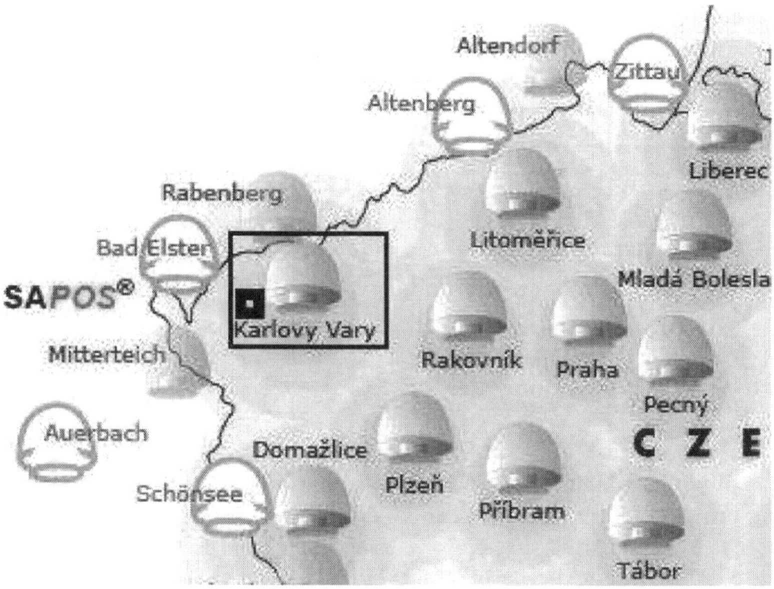

Figure 6. The local network of the GPS reference stations (CZEPOS) in the Czech Republic.

2.3 Data sources from meteorological stations

Data including wind speed, wind direction and atmospheric pressure consisted of time series measurements taken at local meteorological stations. There are two meteorological stations located in the area of the surface coal mine, and a third station which also includes soil humidity measurements is located in the neighbouring residential zone. The data are archived at regular intervals. A selected time period is shown in Figure 7. The wind rose illustrates the predominant wind directions causing increased dust transport to the residential zone. The data processing is based on observations from all the local meteorological stations, which showed similar results for the selected time period.

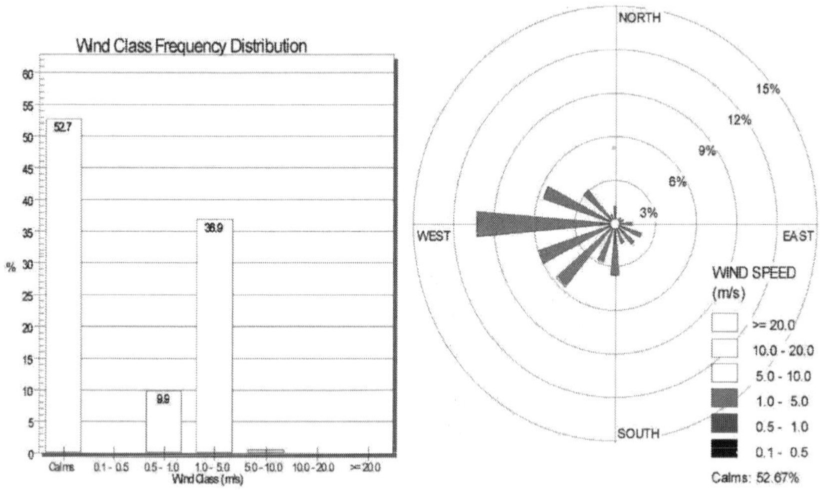

Figure 7. The wind class frequency distribution and the wind rose based on observations from the local meteorological stations (processed by the WRPLOT View software).

2.4 Emission sources

Dust from surface mines is emitted into the atmosphere from a wide range of sources that can be classified as either passive emission sources (temporal coal storage sites, eroded slopes and piles) or active emission sources (sorting sites, excavators, transport routes and other mining equipment). The term dust is non-specific with respect to the size, shape and chemical properties of the particles. Dust is mostly formed by turbulent wind action accompanying the release of gaseous emissions. The concentration of particles in the atmosphere can usually range from a few micrograms to hundreds of micrograms per cubic metre. There have been a number of studies assessing the risk of dust as a health issue for mine workers and surrounding natural environments [8].

In this study we focused on one dominant emission source, the temporary coal storage and sorting site, which was identified by the classification of satellite images or aerial images. In order to more precisely determine its location and shape, terrain measurements with a GPS were used to create a map layer in the framework of the GIS project. Figure 8 shows the area around the temporary storage and sorting site, and the inset shows the current status, with mining installations and temporary stockpiles.

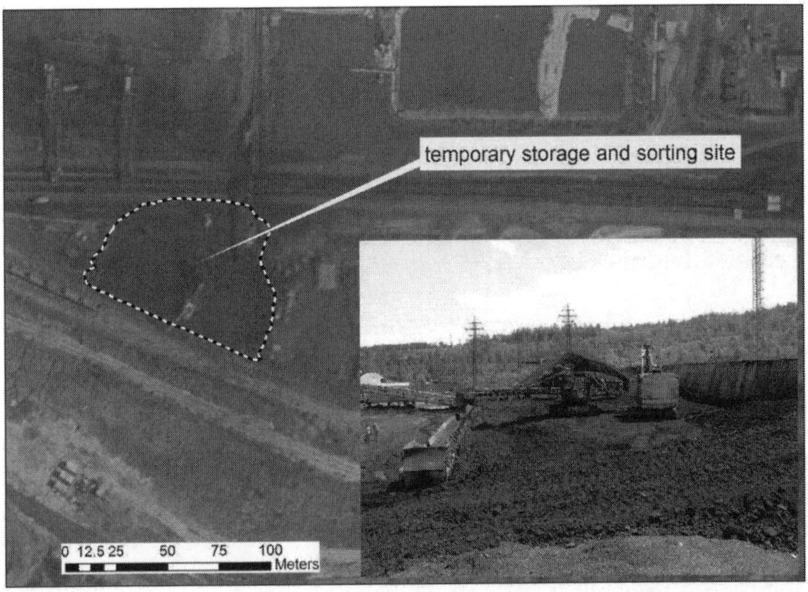

Figure 8. The temporary storage and sorting site located with GPS measurements. The attached image illustrates the actual status of the site.

2.5 Other data sources

Data sources for the DTM are complemented by contour lines from existing digital map sources at a scale of 1:10,000. These maps do not just cover the surface mining area, but extend the DTM to the larger area of the surface mine and its neighbouring zones. Other thematic map sources are Internet based geographic data. For this surface mine, located in the western part of the Czech Republic, Internet mapping servers based on ESRI technology

(http://geoprotal.cenia.cz and http://mapy.kr-karlovarsky.cz) were used to add thematic map layers. In addition to the laser scanning, the local area of the temporary storage site was extended by spatial data obtained from a mining company mapping survey (transport routes, locations of the mining installations, and a set of 3D points).

Satellite images can display larger surface emission sources such as whole surface mine regions. In the case of passive optical sensors, the natural energy detected is emitted or reflected by the objects themselves or by the surrounding area being observed. Then, the spectral, spatial and temporal properties of the gathered information are analysed by the remote sensing software tools. In addition to the natural energy of the visible electromagnetic light spectrum, satellites (LANDSAT, ASTER, SPOT) can detect infrared light that human eyes cannot perceive. Bands in the infrared spectral range can be used for the discrimination of geologic rock types, soil boundaries and soil moisture content which affect the rate of surface emission sources. Relationships with known features on the ground (bare land, vegetation, urban, forest) can be found using multispectral classification. In this study, an orthorectified image from LANDSAT 7 ETM+ (192/25, RMS = 0.63 pixel, approximately 19 meters) was used to display the entire area including the surface mine and residential zones in the framework of the visualization phase.

2.6 The GIS approach

Spatio-temporal modelling including spatial analysis and extensions for the analysis of time series offers a wide range of advanced functions for data processing, visualization and digital mapping outputs [9-11], and thus can support environmental planning and policy making. Direct access to large volumes of data allows more complex information to be used in decision-making processes, while the ability to reach a broad audience and to generate a large number of options, scenario analyses and forecasts can help to more efficiently solve environmental problems [12]. In the case of dust transport, insight into wind flows together with the DTM, aerial images and pre-processed data from meteorological stations and existing thematic map layers helps in the exploration of the mechanisms of dust dispersion and deposition. This approach focused on the integration of spatio-temporal data and the workflow of data processing is illustrated in Figure 9.

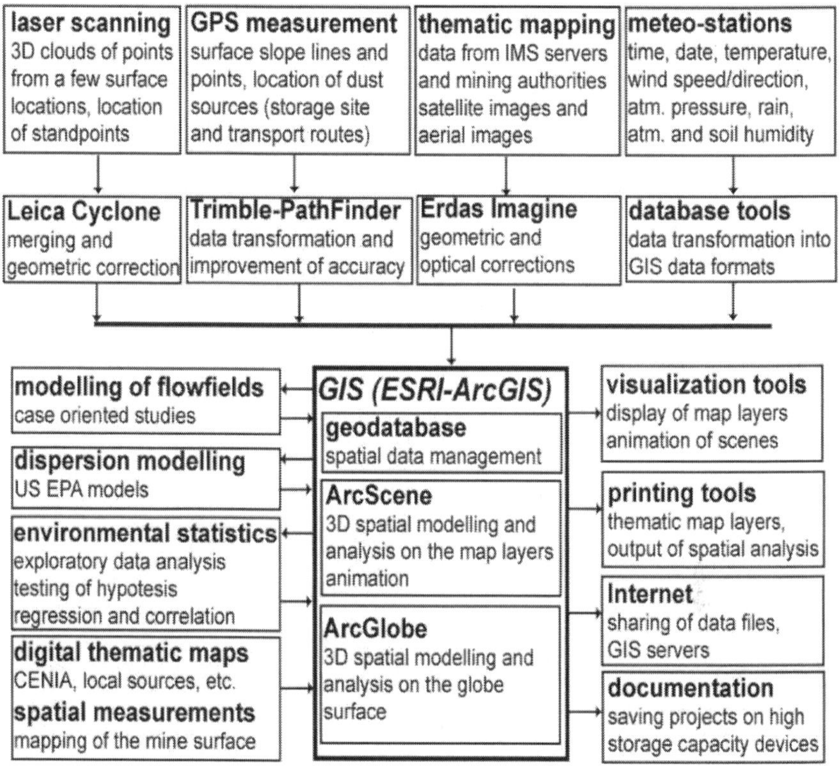

Figure 9. The general workflow of data processing in the GIS environment.

The GIS project based on spatio-temporal data serves as the main data storage for numerical modelling. Data inputs for the numerical simulations contain the DTM and time series from the meteorological stations. Data outputs from numerical simulations are represented by the three dimensional domain over the DTM. The domain forms layers, which in turn contain a set of nodes. The attributes of each node include the coordinates in 3D space, the velocity vector, atmospheric pressure and the concentration of the passive pollutant. In order to import data into the GIS project, the post processing phase is focused on georeferencing of the data domain, transformation of the internal data and import into the GIS database.

2.7 The modelling approach

The dispersion of dust particles near an emission source can be described by mathematical equations based on physical principles [13].

However, in the actual environment with a huge number of various surface objects (i.e., the temporary storage site, mining installations and buildings) and other factors (climate, soil conditions), the models reflect only basic phenomena and have to be improved through many corrections depending on the selected mining area and field measurements [14, 15].

Due to this approximate character of the results from mathematical modelling, an alternative method is to use wind tunnels for physical modelling. These types of experiments can be performed on down-scaled models of the studied area, with simulated flow conditions similar to the conditions observed in the atmosphere. In order to validate flow fields and pollution dispersion in simplified 2-D and 3-D space, field measurements are necessary to explore all the spatio-temporal characteristics which depend on surface objects [16].

In this paper, wind flows and pollution dispersion in atmospheric boundary layers were simulated through numerical models based on Reynolds averaged Navier-Stokes equations (RANS) in non-conservative form:

$$u_x + v_y + w_z = 0 \tag{1}$$

$$V_t + uV_x + vV_y + wV_z = -\frac{\nabla p}{\rho} + [KV_x]_x + [KV_y]_y + [KV_z]_z, \tag{2}$$

where $V = col(u,v,w)$ is the velocity vector, p pressure, ρ density.

The transport equations for concentrations of passive pollutants are:

$$C_t^i + uC_x^i + vC_y^i + wC_z^i = \left[K\frac{C_x^i}{\sigma_{C^i}}\right]_x + \left[K\frac{C_y^i}{\sigma_{C^i}}\right]_y + \left[K\frac{C_z^i}{\sigma_{C^i}}\right]_z, \tag{3}$$

where C^i is the concentration of the i_{th} pollutant and σ denotes the turbulent Prandtl's number. The turbulence model is based on the Boussinesq hypothesis of the turbulent diffusion coefficient:

$$K = \nu + \nu_T \tag{4}$$

that is represented by the sum of molecular and eddy viscosity. Thus, the final implemented algebraic turbulence model is in the form:

$$K = \nu + l^2\sqrt{\left[\left(\frac{\partial u}{\partial z}\right)^2 + \left(\frac{\partial v}{\partial z}\right)^2\right]},\ \text{where}\ l = \frac{\kappa(z+z_0)}{1+\kappa\frac{z+z_0}{l_\infty}},\ l_\infty = \frac{27|V_G|10^{-5}}{f_c}. \tag{5}$$

The parameter $f_c = 0.00011\ ms$ denotes the Coriolis parameter. The parameter V_G is the geostrophic wind. The numerical simulation was carried out with a semi-implicit finite-difference scheme [17] extended by a number of other simulation tools focused on pollution dispersion in 3D atmospheric boundary layers [18].

The outputs from numerical simulations were formed in the 3D domain and divided into layers with a set of spatial points. These layers are located over the area of interest. An example of data exchange by shared files is illustrated in Figure 10. Initially, the DTM was exported into an ASCII grid as the input file for the numerical simulation of wind flows and pollutant concentrations (cell size: 1 meter; national coordinate system in the Czech Republic: S-JTSK). The outputs from numerical modelling were then backward imported into the GIS database. The text file contained 3D points (X, Y, Z coordinates) and their attributes: atmospheric pressure (p), wind speeds in the x, y and z direction (u, v, w), and dust concentration (c). Finally, the GIS database (ESRI's geodatabase in ArcGIS) contained 21 point layers from the imported 3D domain located over the area of interest. The outputs from numerical simulation (values of variables: p, u, v, w and c) were then transferred into attribute tables for subsequent processing and visualization.

2.8 Visualization of the dust transport

There are various methods for visualization that depend on available computer simulation tools. Existing visualization tools are mostly included with simulation packages as extensions or programming libraries [19]. Some visualization tools are included directly into the user interface [20]. Many methods are implemented in the GISs that can usually employ a two-dimensional framework included into the map layers. But, wind flows and dust transport require a more dynamic way of data processing and visualization, because the processes vary through three-dimensional space and through time. In spite of the fact that conventional GISs can handle the third dimension and time as attributes of basic geographical objects, other standalone software applications are needed to offer more advanced visualization functions. In many cases, existing GISs are extended by modules that provide visualization toolkits for 3D spatial data and time-varying data [21]. In this instance, ArcScene in ArcGIS was used for the visualization of atmospheric pressure, wind flows and dust concentrations over the area of interest. As an example, Figure 11 illustrates the 3D visualization of the dust concentration in one selected layer. The image is based on spatial data exported into the graphic files that allow only a limited demonstration of full functionality.

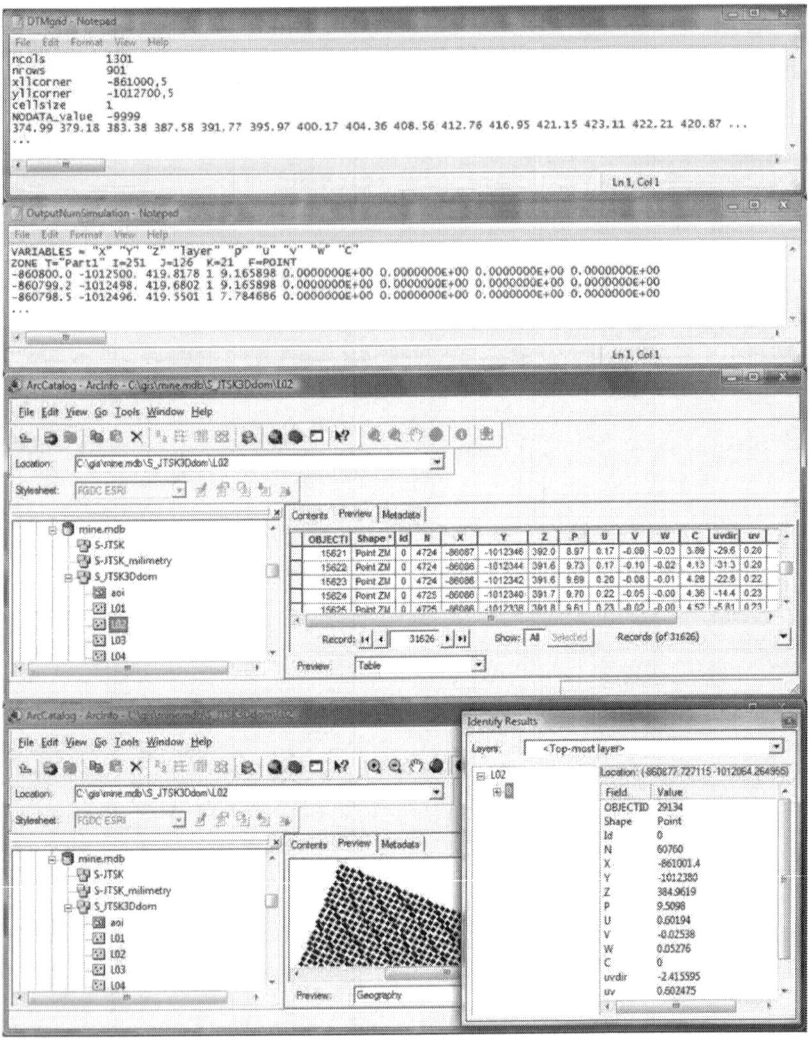

Figure 10. The data exchange by shared files between the GIS and numerical simulation **(a)** An example of the DTM exported into the ASCII grid for numerical simulation. **(b)** An example of the outputs from numerical simulation represented by 3D points (*X*, *Y*, *Z*coordinates) and their attributes: atmospheric pressure (*p*), wind speeds in the *x*, *y*, *z* direction (*u*, *v*, *w*), and dust concentration (*c*). **(c)** An example of the geodatabase (*mine.mdb*) in ArcGIS containing selection of the point layer with the attribute table based on data from the numerical simulation (illustrated on the left). **(d)** An example of the geodatabase (*mine.mdb*) in ArcGIS containing selection of a point.

Figure 11. (**a**) An example of the visualization of dust concentration in the selected horizontal layer 5 meters over the DTM (extended by the simulation grid and exported from ArcScene). **(b)** A more detailed view of the visualization.

The 3D domain transferred into the GIS contains 21 layers with a height increment of 1 meter for the 15 bottom layers, and a height increment of

100 meters for the 6 upper layers. For example, the 2 bottom layers with the prediction map of dust concentration over the area of interest are shown in Figure 12. The first bottom layer (L1) is 1 meter over the DTM, and the second layer (L2) is 2 meters over the DTM. In order to study wind erosion processes over the coal stockpiles in a more precise way, Figure 13 illustrates a detailed view of the first bottom layer over the temporary storage and sorting site. The arrows show the wind direction and wind speed over the stockpiles. Vertical drift is illustrated by the prediction map in the background.

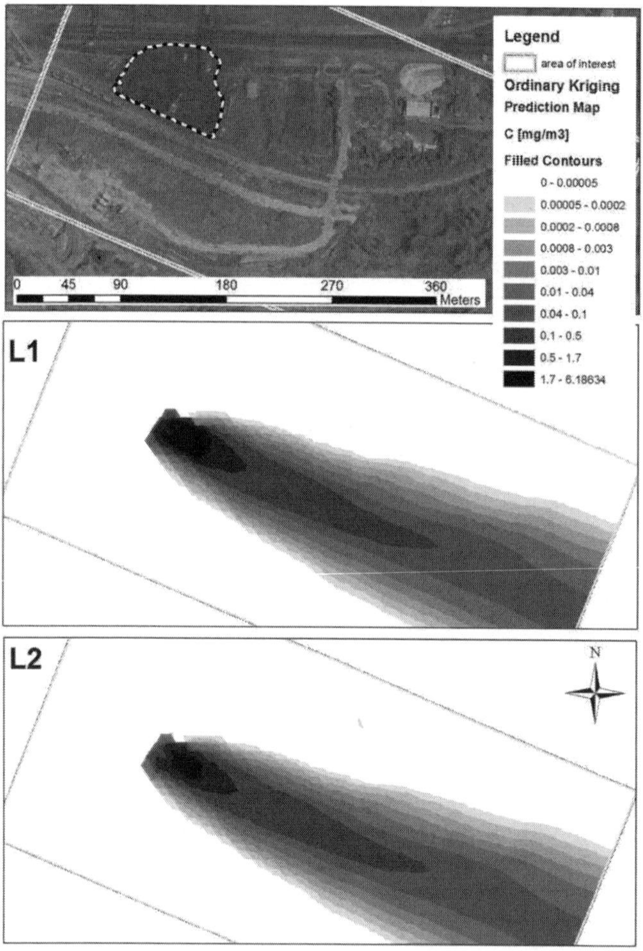

Figure 12. An aerial image of the area of interest followed by the two bottom layers (L1, L2) with the prediction maps of dust concentration. The prediction maps are based on spatial interpolation (ordinary kriging) of the data transferred from the numerical simulation.

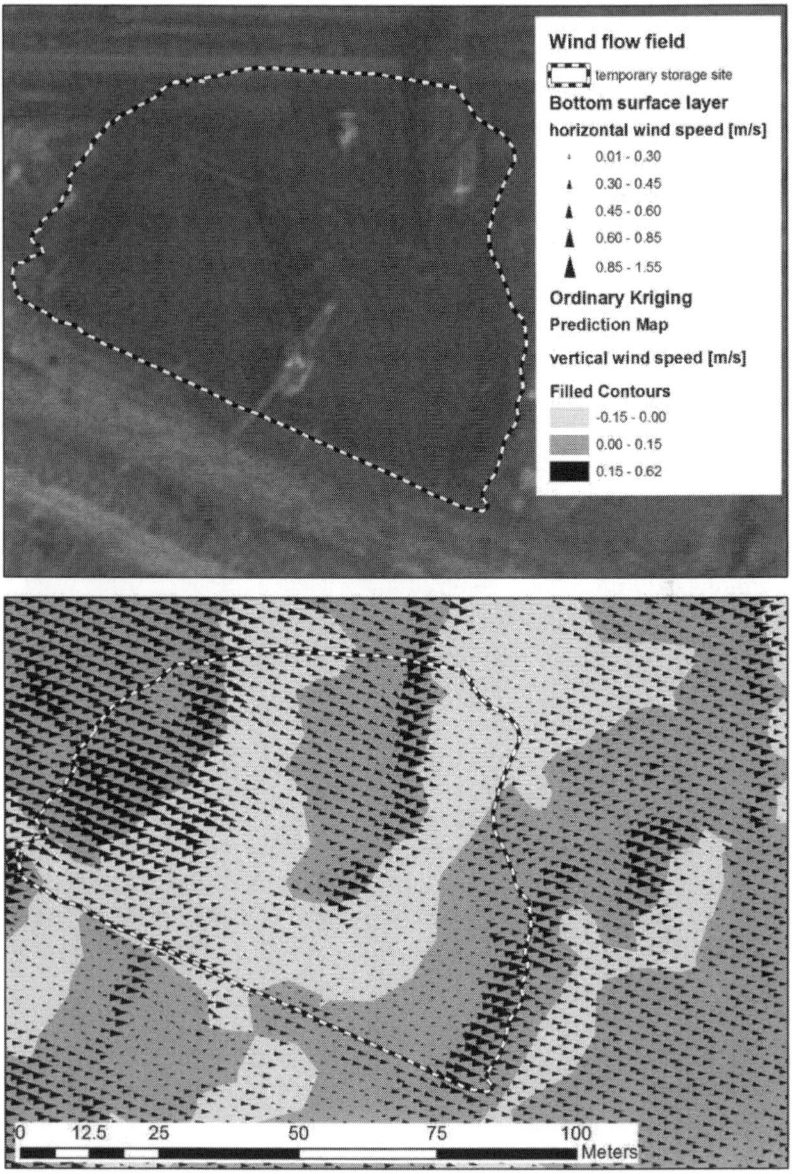

Figure 13. An aerial image of the temporary storage site complemented by the wind flows in the first bottom layer over the storage site (approximately 1 meter over the stockpiles). The arrows show the horizontal speed and direction; the filled line contours on the background show the vertical speed.

3. DISCUSSION AND SUGGESTED FUTURE RESEARCH

This study describes the extension of the capabilities for spatio-temporal analysis, with an emphasis on physically based dust transport and coupling with GIS. This more precise numerical simulation of dust transport requires a more complex DTM construction. This DTM is extended by additional surface objects, some spatial data sources and field measurements, which are then used for construction in the GIS environment. At the same time, various results can be compared depending on accuracy and capture efficiency.

The use of 3D laser scanners enables more accurate continuous mapping of smaller areas for the DTM, but multiple scans from different locations are required to obtain complete data sources from parts that are obscured. Also, more powerful software tools are needed to bring the clouds of spatial points into a common reference system and to eliminate incorrect spatial points, originating for example from vegetation, pylons and mining equipment. In general, a reduced data set is first used for integration into the GIS database in order to create the DTM. Then, the DTM based on point clouds captured with average grid 0.2 meters are a necessary spatial data source for representing the vertical slopes of the surface mines.

GPS measurements with a final accuracy of 0.5 m after the postprocessing phase complement the data from 3D laser scanners. Besides capturing other 3D points and 3D lines for the DTM, the GPS was used for the localization of other surface objects for thematic mapping (i.e. boundaries of the temporary storage site with its coal stockpiles, actual transport routes, coordinates of meteorological stations), and also to complement existing thematic maps of the neighbouring residential zones.

Other 3D spatial data, mostly based on 1:10,000 thematic maps, are not used for the DTM which supports the numerical simulations, but are dedicated to the overall visualization of larger areas extended by data from numerical modelling and spatial analyses. Also, the aerial images and satellite images from Landsat 7 ETM+ are used to more realistically display the area of interest.

Pre-processed data from the DTM and time series of the wind speed and direction from the local meteorological stations are used as data inputs into the numerical simulation based on RANS equations. The simulation output represents a 3D domain of nodes with defined velocity vectors, atmospheric pressure values and dust concentrations. The spatial projection of the 3D domain over the DTM enables the identification of areas of higher wind speed with vertical drift near the stockpiles, where dust emissions are most likely to occur. In addition to the spatial linkage of the simulation outputs with the DTM, this integration of thematic map layers and aerial images in the GIS environment allows new ways to interpret results. Thus, more precise terrain mapping of the surface by GPS and classification of the aerial images can help to improve estimates of dust emission and the modelling of dust transport. Until recently, a number of studies focused on flows over simplified objects such as stockpiles or man-made barriers have been carried out using numerical modelling or simulations in wind tunnels [22, 23]. However, it is difficult to apply these partial results in the general scale of wind flow spatio-temporal modelling over the DTM in real conditions. Thus, to improve model accuracy, more complex measurements of emission processes which depend on meteorological conditions and soil properties are needed to test other effects and to propose more efficient procedures for decreasing dust emissions and deposition.

In order to implement spatio-temporal modelling in the framework of dust transport risk assessment, a standalone application in the GIS environment needs to be developed that can share a wide range of existing spatial data and provide on-line numerical simulations, spatial analysis of dust concentrations and final visualization. Various case studies could then be tested by mining authorities and local agencies, in order to minimize the actual coal emissions and transport.

4. CONCLUSIONS

The presented spatio-temporal modelling of dust transport over a surface mining area and neighbouring residential zones offers an excellent opportunity for an exploration into dust emission, transport and deposition coupled with data from remote sensors like 3D laser scanning and aerial images. The created GIS project is able to analyse a wide range of data in order to assist in decision-making processes for decreasing dust deposition in residential zones. Various configurations of man-made barriers as well as different flow conditions can be tested to find the most appropriate solutions.

ACKNOWLEDGMENTS

The presented research was carried out in the framework of the projects supported by Academy of Sciences of the Czech Republic (AVCR 1ET400760405), and Ministry of Education of the Czech Republic (MSM 6840770010). The GIS project has been processed with ArcGIS and ERDAS Imagine in the GIS Laboratory at the Faculty of Science, Charles University in Prague.

REFERENCES AND NOTES

1. Shaw, W.J.; Allwine, K.J.; Bradley, G.F.; Frederick, C.R.; Rishel, J.P.; Chapman, E.G. An evaluation of the wind erosion module in DUSTRAN. Atmospheric Environment 2008, 42, 1907–1921.
2. Webb, N.P.; McGowan, H.A.; Phinn, S.R.; McTainsh, G.H. AUSLEM (AUStralian Land Erodibility Model): A tool for identifying wind erosion hazard in Australia. Geomorphology 2006, 78, 179–200.
3. Chakraborty, M.K.; Ahmad, M.; Singh, R.S.; Pal, D.; Bandopadhyay, C.; Chaulya, S.K. Determination of the emission rate from various opencast mining operations. Environmental Modelling & Software2002, 17, 467–480.
4. Maidment, D.R. Environmental modeling within GIS. In GIS and Environmental Modeling: Progress and Research Issues; Goodchild, M.R., Steyaert, L.T., Parks, B.O., et al, Eds.; GIS World Books: Fort Collins, CO, USA, 1996; pp. 315–323.
5. Matejicek, L. Spatial modelling of air pollution in urban areas with GIS: a case study on integrated database development. Advances in Geosciences 2005, 4, 63–68.
6. Fedra, K. Distributed models and embedded GIS. In GIS and Environmental Modeling: Progress and Research Issues; Goodchild, M.R., Steyaert, L.T., Parks, B.O., et al, Eds.; GIS World Books: Fort Collins, CO, USA, 1996; pp. 413–417.
7. Vosselman, G.; Kessels, P.; Gorte, B. The utilisation of airborne laser scanning for mapping.International Journal of Applied Earth Observation and Geoinformation 2005, 6, 177–186.
8. Mower, B. A multiple source approach to acute human health risk assessment. Waste Management1998, 18, 377–384.
9. Mitchell, A. GIS Analysis: Geographic Patterns & Relationships; ESRI Press: Redlands, CA, USA, 1999; Volume 1, pp. 9–36.
10. Mitchell, A. GIS Analysis: Spatial Measurements & Statistics; ESRI Press: Redlands, CA, USA, 2005; Volume 2, pp. 1–70.
11. Longley, P.A.; Goodchild, M.F.; Maguire, D.J.; Rhind, D.W. Geographic Information Systems and Science; John Wiley: New York, NY, USA, 2001; pp. 183–276.
12. Parker, P.; Letcher, R.; Jakeman, A.; Beck, M.B.; Harris, G.; et al. Progress in integrated assessment and modelling. Environmental Modelling & Software 2002, 17, 209–217.
13. Meroney, R.; Bienkiewicz, B. Computational Wind Engineering 2; Elsevier: Amsterdam, Holland, 1997; pp. 375–508.

14. Skote, M.; Sandberg, M.; Westerberg, U.; Claesson, L.; Johansson, A.V. Numerical and experimental studies of wind environment in an urban morphology. Atmospheric Environment 2005, 39, 6147–6158.

15. Gromke, C.; Ruck, B. Influence of trees on the dispersion of pollutants in an urban street canyon-Experimental investigation of the flow and concentration field. Atmospheric Environment 2007, 41, 3287–3302.

16. Badr, T.; Harion, J.T. Numerical modeling of flow over stockpiles: Implication on dust emissions.Atmospheric Environment 2005, 39, 5576–5584.

17. Kozel, K.; Bodnar, T.; Gulikova, E.; Pisa, V. Numerical simulation of flow and pollution dispersion over obstacles in complex terrain. In Colloquium Fluid Dynamics 2007; Jonas, P., Uruba, V., Eds.; Institute of Thermomechanics, Academy of Sciences: Prague, Czech Republic, 2007; pp. 43–44.

18. Benes, L.; Bodnar, T.; Fraunie, P.; Kozel, K. Numerical modelling of pollution dispersion in 3D atmospheric boundary layers. In Air Pollution and Simulation; Sportisse, B., Ed.; Springer Verlag: Berlin, Germany, 2002; pp. 69–78.

19. Littlefield, B.L.; Hanselman, D.C. Mastering Matlab 7; Prentice-Hall: New York, NY, USA, 2004; pp. 1–38.

20. Dabney, J.B.; Harman, T.L. Mastering Simulink; Prentice-Hall: New York, NY, USA, 2003; pp. 1–40.

21. D□llner, J.; Hinrichs, K. An object-oriented approach for integrating 3D visualization systems and GIS.Computer & Geosciences 2000, 26, 67–76.

22. Lu, H.; Shao, Y. Toward quantitative prediction of dust storms: an integrated wind erosion modelling system and its application. Environmental Modelling & Software 2001, 16, 233–249.

23. Craig, K.J.; de Kock, D.J.; Snyman, J.A. Minimizing the effect of automotive pollution in urban geometry using mathematical optimization. Atmospheric Environment 2001, 35, 579–587.

6

UML PROFILE FOR MINING PROCESS: SUPPORTING MODELING AND SIMULATION BASED ON METAMODELS OF ACTIVITY DIAGRAM

Andrea Giubergia,[1] Daniel Riesco,[2] Verónica Gil-Costa,[1] and Marcela Printista[2]

[1]Mine Department, Chacabuco 569, National University of San Luis, 5700 San Luis, Argentina

[2]Informatic Department, Ejército de los Andes 950, National University of San Luis, 5700 San Luis, Argentina

ABSTRACT

An UML profile describes lightweight extension mechanism to the UML by defining custom stereotypes, tagged values, and constraints. They are used to adapt UML metamodel to different platforms and domains. In this paper we present an UML profile for models supporting event driving simulation. In particular, we use the Arena simulation tool and we focus on the mining process domain. Profiles provide an easy way to obtain well-defined specifications, regulated by the Object Management Group (OMG). They can be used as a presimulation technique to obtain solid models for the mining industry. In this work we present a new profile to extend the UML metamodel; in particular we focus on the activity diagram. This extended model is applied to an industry problem involving loading and transportation of minerals in the field of mining process.

1. INTRODUCTION

The literature on formal use of Unified Modeling Language (UML) [1] is huge and growing rapidly. This tool becomes a modeling standard in 1990 to follow what changed the world of software [2]. UML makes it possible to model any system from different perspectives. Modeling a system from different perspectives allows to clearly picking the vision of what you want to do. This is because UML is a language that holds its own specific rules, semantic, and syntax.

UML can be used as documentation of code, but it is also intended as a means of specifying a system. Model Driven Architecture (MDA) [3] calls for complete systems to be generated automatically from UML models. If the language is not precisely defined, the generated system may not be what the model creators intended.

UML has an abstract syntax which is processed by model transformation and code generation. The relation in UML between concrete diagrammatic syntax and the abstract syntax it represents is complicated enough to be a potential source of error. Precisely defining this relationship could simplify the creation of graphical model editors and facilitate animations [4] and reverse engineering. The definition should clearly delineate concrete syntax, abstract syntax, and semantics, and it should also specify the relationships between these parts.

UML defines several types of diagrams to view the dynamic aspects of a system. One of these diagram types is the activity diagram [5] which is used in this work to document the workflow in a system. UML Diagrams are suitable for system analysis, design, and development. However, UML and its diagrams cannot accurately specify concepts for particular domains. That is why UML embodies the concept of profiles. Profiles are defined by the Object Management Group (OMG) in its specification [1], as a predefined set of stereotypes, tagged values, constraints (by using Object Constraint Language (OCL)), and notation icons that collectively specialize and tailor UML to a specific domain or process. When these elements are introduced the model can be clearly visualized and software developers can improve communication and establish a common vocabulary. Also profiles allow adding information to the model to transform it to other models. Object Constraint Language (OCL) constraints are semantic restrictions added to UML elements. The advantage of profiles is that most UML tools can easily apply them. When using profiles it is not necessary to define neither a special notation nor special tools (UML tool is used).

The UML metamodel (which consists of entities and relationships) of the domain is a process-oriented tool. It is graphically structured for the construction of diagrams or flowcharts. These diagrams show the number of steps required by an entity as it moves into the system.

Another powerful tool for system analysis and design is simulation. Its benefits are applicable to almost all kinds of industry. It involves designing a model of a particular system to solve it by means of a numerical technique (algorithm) and the subsequent execution of a series of experiments with the aim of understanding the behavior of such system under certain conditions [6]. The model should be able to reproduce the actual process behavior as accurately as possible. There are many simulation softwares of general purpose [7–10]; in this work we focus on the Arena [10] simulation because it has become the market-leading discrete event simulation software.

To reconcile the two ways of approaching the same subject, namely, UML and simulation, in this work we propose to model a system of loading and transportation of material in the field of the mining industry. In particular, we propose to extend the UML metamodel using profiles. The objective is to adapt the UML metamodel to be able to represent the basic modules of Arena.

The advantage of the proposed profile is that (a) it allows reusing the stereotypes created for the Arena simulation software. Moreover, those stereotypes can be applied to any other process-oriented simulation software. We can also use them or integrate them as part of any other process-oriented simulation software.

This paper is organized as follows. Section 2 presents related work. The Arena simulation software is briefly described in Section 3. Section 4 presents our mining process case of study. Section 5 presents the proposed profile implementing a simulation environment and its OCL constraints. Conclusions are included in Section 6.

2. RELATED WORK

Modeling and simulation techniques provide the possibility of studying new strategies and to predict the effect of new policies, new designs, and new strategies which would otherwise be too expensive or even impossible to implement and test on real cases. Both the UML modeling and the simulation approach model from two different communities (or point of view) but they can be used for similar purposes.

Some recent methodologies and techniques for modeling and simulation have been presented in [11–13]. In particular, the authors in [11] use genetic algorithms to improve component analysis. This technique is applied to simulated data collected from the Tennessee Eastman chemical plant. The work in [12] evaluates basic data-driven methods for process monitoring and fault diagnosis. Also in this case the benchmark of Tennessee Eastman (TE) process is utilized to illustrate the efficiencies of the discussed methods. Finally, in [13] the authors present a fault tolerant control architecture which can be reconfigured online. The proposed scheme is evaluated with the TE benchmark model.

In this work we focus on UML modeling and simulation techniques which are widely used in systems engineering. Although these two methodologies have evolved separately, there are some works [14–17] that integrate modeling tools with specific simulation software.

The works presented in [14, 15] show how the UML-Arena combination is used to create a model. Then this model is reused into the simulation environment of the Arena software. In [14] the authors describe the use of activity diagrams to automatically generate simulation models and propose the use of an algorithm as a solution to automatically transform UML models (only activity diagrams) into simulation models. Similarly, the authors in [15] show the use of UML activity diagrams as the basis for building simulation models, which are then run on the Arena simulation software.

In other words, these studies show a methodology to transform activity diagrams into simulation models for Arena. However, they do not take advantage of the UML profiles. A profile created for a specific situation can be reused into other environments with similar characteristics, thus saving time and increasing the versatility of the profile created.

As explained in [18], OMG defines two possible approaches for defining domain specific languages. In the first one, a new alternative language is defined [19]. A new tailor-made language will produce suitable specific notation that will match the concepts of a specific application domain. However, as the new language does not respect UML semantics, it will not allow the use of commercial UML tools for drawing diagrams, generating code, and so forth [18]. The second approach (named Profiles) uses the UML metamodel respecting the original semantic of the UML elements (classes, attributes, etc.). But new constrains are added to the definition and to the relationships of those elements.

The works in [20–22] show the versatility of profiles to define different environments (application domains), which extend the UML specification (not only the syntax but also the semantic through formal OCL constraints). In particular, the work in [20] shows the way for defining design patterns with profile, proposing architecture in levels. It shows how the definition of a profile for a particular pattern, and how an UML tool can be enough to introduce profile for patterns. It analyzes the advantages of using profiles to define, document, and visualize design patterns.

The work in [22] shows the definition of the C&K metrics applied to Aspect Oriented Design (AOD) using UML profiles. A new definition of metrics using profiles without modifying the OMG metamodels is specified. An OCL formal specification is developed. The defined profile allows applying methodologies which use AOD and allows measuring the design. The calculation of the resources is an activity that can improve the software development. The engineers can import this profile to any UML tool and measure the quality of its AOD. They do not need to build a specific tool for AOD. They can use any UML tools and thereby improve the quality of aspect oriented developments.

In [21], the creation of components of the Java EE 6 business platform from technical business processes modeled with Business Process Model Notation (BPMN) 2.0 is proposed. This generation was achieved by performing three transformations in the context of Directed Model Architecture (MDA). First, a technical model BPMN 2.0 is transformed into an UML class model. Then the class model is transformed into a model with Enterprise Edition (Java EE) profiles. Finally the last generated model is transformed through Meta Object Facility Script (MOFScript) into Java EE components. The transformations are performed with Query View Transformations (QVT) Relations and MOFScript. This work contributes with transformations between profiles generated with Java EE business components related to business processes, so it helps in improving the development productivity and in reducing design errors.

In [16], UML is proposed as an effective structure for building simulation models of hybrid manufacturing systems, where machines communicate through a network or using buffers. The model is implemented with the MATLAB language [23]. UML is used for modeling, but UML profiles are not considered.

The work in [24] presents research results related to the main steps of the modeling and simulation process (e.g., design of simulation experiments), as well as to the application of simulation for solving different problems, inherent to the operation of complex systems (e.g., analysis, optimization, and management). In particular, the authors describe a tool which integrates simulation methodologies that incorporate simulation with other scientific approaches for the analysis, optimization, and management of complex systems. This tool allows for transforming UML models into a simulation environment; in particular the Arena simulation environment is used.

UML has also been used in collaboration with other simulation softwares like DEVS [25] or Petri Nets. In [26], the authors present UML diagrams that are extended with stochastic attributes. This provides possibilities for further simulation of a developed model using a simulation tool called DEVS [25]. The authors generate UML models using use case and activity diagrams. On the other hand, the work in [17] proposes a methodology for transferring knowledge to students. The authors propose to combine the use of Petri Net and UML to model a business process as a system of discrete events. None of these papers create UML profiles.

In this paper we focus on the use of UML activity diagrams such as a presimulation techniques, because they allow capturing dynamic aspects of the behavior of a system. We extend the UML metamodel by defining profiles for a particular domain. Unlike previous work, like described in [14, 15], our work ensures that the UML metamodel is not modified and it also meets all the semantic. Moreover, our proposed profile is specified with the formality given by the OCL [27] language, allowing (1) to reuse the proposed stereotypes with any other simulation software and (2) to reuse the simulation model with any tool that includes profiles.

3. SIMULATION WITH ARENA

Modeling and simulation provide the basis for the efficient solving of various problems related to the operation of complex systems like analysis, optimization, and management and industry problems like mining process, and so forth. Simulation is considered to be one of the most effective technologies for the analysis and planning of logistics systems.

Using simulation, you can include randomness through properly identified probability distributions taken directly from study data. For example, in this work we consider the loading and transportation of material in the field of the mining industry. In this case, there may be intersections on roads or narrow paths that only allow a truck to move at a time. Simulation also allows including the analysis and evaluation of the system when failures occur.

Other techniques such as spreadsheet analysis or linear, goal, and dynamic programming are useful to maximize or minimize a single element (e.g., cost, utilization, or wait time). But these techniques limit the analysis to only one element, often at the expense of secondary goals. They do not allow randomness. In particular, spreadsheet analysis forces to use the average time and will not be able to accurately capture the variability that exists in reality.

Arena simulation software [10] is a general propose simulation tool enabling the construction of models over a series of modules or basic components organized hierarchically. The Arena simulation software has high level of modeling supporting graphical design. It also includes a lower level of modeling including specific details as arrival times, service times, scheduling of processes, and so forth.

A model is developed using modules that are part of the basic processes. In Arena, modules are the flowchart and data objects that define the process to be simulated. All information required to simulate a process is stored in modules. The dynamics associated with the processes can be viewed as nodes in a network by which entities circulate causing a change in the system state. The entities with attributes and variables compete for the services provided by the resources. Entities are items (like trucks, mineral, etc.) that are being served or produced.

Figure 1 shows a simple model build with Arena. The CREATE module is the starting point for entities in a simulation model. Entities are created using a schedule or based on a time between arrivals. Entities then leave the module to begin processing through the system. The entity type is specified in this module. The PROCESS modules are intended as the main processing method in the simulation. They include the resource by which entities compete. A resource retained by an entity must be released at some point in the model. Otherwise, a deadlock can occur. The DECIDE module allows for decision-making processes in the system. Finally, the DISPOSE modules are ending point for entities in a simulation model.

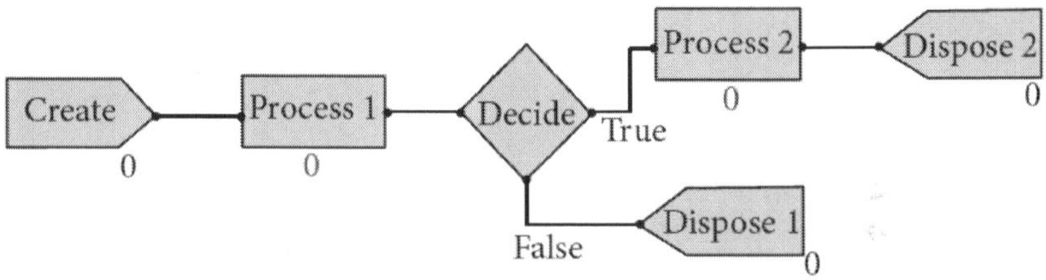

Figure 1: Basic modules used in Arena.

After the model is built, we can run simulations to obtain different metrics and statistics like resources utilization, waiting times, and so forth.

4. CASE STUDY: MINING PROCESS

A real system model is fundamental when making crucial decisions. A system model offers the possibility of planning with less error. But more importantly, it is possible to investigate new operating procedures that can be used repeatedly without interfering in the operational and daily activities of the system.

The profile proposed in this work is applied to a system of load and transportation of minerals. During the mining process, once the rock or mineral has been removed from its extraction site, it needs to be transported immediately to its final destination. Generally, the transport system consists of a set of trucks.

Several factors are considered for the selection of trucks, the most important factor is the capacity of the vehicle in the mine. To meet daily production requirements is necessary: (a) a good performance of the equipment, that is, to achieve the perfect match "shovel-truck" to obtain maximum extraction and transportation of minerals and (b) the road conditions should be well maintained.

The tasks performed in the daily cycle of a mining process [28, 29] start with the definition of a drilling mesh (according to height, quality, and quantity of the field to be explode). Then, holes are drilled into the field. Then the explosives are introduced into the hole for blasting. The aim is to detach the rock out of the field and crush it. Then blasted material is loaded on the trucks and transported to a discharge area.

4.1. Mining Process Simulation with Arena

In this section we explain how to perform the design of load and transportation of minerals in the field of mining process using the Arena simulation tool. However, we emphasize that the focus of this paper is not the simulation design/evaluation of the mining operations but the profile design for discrete event simulations applied to the field of mining projects.

We model a transportation system for a limestone quarry. The trucks move between their loading points to their discharge points using default routes. The cycle begins with the release of trucks (≪Create≫) to operating fronts. Trucks are distributed among three different zones (using the module ≪Decide≫) to load the mineral. The distribution percentage of the ≪Decide≫ module is based on the quality and quantity of material. Once in the loading zone (≪Process≫), if the blade (≪Resource≫) is free then the truck is loaded without delay. Otherwise, it should be queued (≪Queue≫) until the resource is released.

When the loading process ends, the truck is directed to the discharge zone (≪Process≫), which may be inside the plant or in the sterile area, according to the specific material being transported. This cycle is repeated according to the scheduled work.

Tables 1 and 2 present time involved in a limestone quarry to move trucks from/to the treatment plant to/from the operating fronts. Trucks unload the mineral into a chute in the treatment plant. At the operating fronts we find shovels to load the trucks. Unloaded trucks move faster than loaded trucks. Then the time required to move unloaded trucks from the treatment plant to the operating fronts is smaller than the time required to move the loaded trucks from the operating fronts to the treatment plant as shown in Table 1. In Table 2 we show that the time to load/unload a truck follows an exponential distribution.

Table 1: Time (minutes) for mineral hauling.

Route	Shovel 1	Shovel 2	Shovel 3
From plant to solvers	3	5	10
From solvers to plant	8	10	20

Table 2: Time (minutes) for load/unload the mineral to/from the truck.

Shovel 1	Shovel 2	Shovel 3	Treatment plant
Exp(6.3)	Exp(6.2)	Exp(6.4)	Exp(5.6)

Figure 2 shows the simulation design implemented with Arena. In this particular case, we evaluate the transportation system of a limestone quarry with tree operation fronts. We set the maximum number of trucks to 5. The number of trucks has been chosen to avoid long queuing times. From each operation front the trucks are loaded with minerals with different "low values," namely, with different quality of the mineral. In each operation front there is a shovel in charge of putting the mineral in the truck. To decide the route for each truck we use an "N-way by chance" ≪Decide≫ module and we set the percentage of each route to 33%.

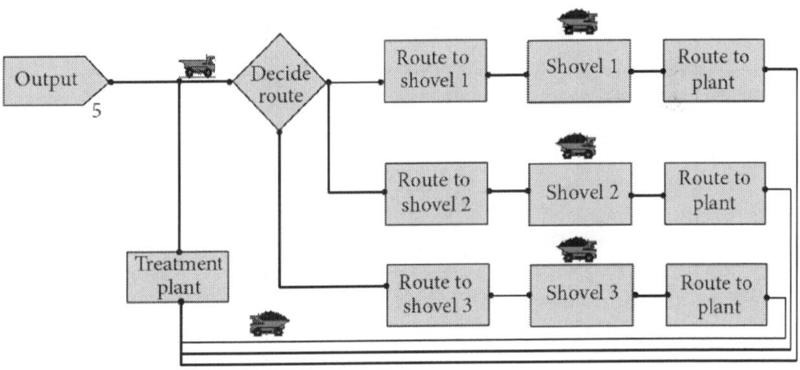

Figure 2. Limestone quarry simulation.

Table 3 shows results obtained after we run the simulation for 20 days (each day has 16 working hours). As we said before the number of trucks is set to avoid long queuing times. Therefore results show that the average number of waiting trucks in each ≪Resource≫ (treatment plant, shovels 1,

2, and 3) is less than 1 and the queuing time in the worst case is less than 1 minute.

Table 3: Simulation results.

Resource	N° waiting	Avg. waiting time	Max. waiting time
Treatment plant	0.81	0.102	0.889
Shovel 1	0.06	0.024	0.729
Shovel 2	0.08	0.032	0.684
Shovel 3	0.07	0.027	0.596

5. UML PROFILE DEFINITION FOR MINING PROCESS SIMULATION

UML uses a mechanism called profile [1] to adapt a model to a particular domain. A profile includes three elements: stereotypes, tagged values, and constraints. Stereotypes extend the vocabulary of UML and it is possible to associate tagged values (attributes associated with extended elements) and restrictions [20, 22]. These extension mechanisms allow to adapt a particular domain to an existing metamodel.

The main builder of the profile is the stereotype, indicated as ≪stereotype≫ [1]. It has the same structure or elements (attributes, associations, and operations) of the UML metamodel. So, it is not allowed to modify the semantics, structure, and original concepts [30] of the standard UML.

The proposed stereotypes are related to the components of the activity diagram [5], because they easily adapt themselves to characterize the simulation model [31, 32]. This diagram already has established appropriate notations and concepts showing the steps (activities), decision points, and bifurcations that occur in a transaction so that makes it easier to display. In particular in this work we used the Rational Software Architect (RSA) tool [33] which has the advantage of supporting OCL restrictions. But any tool supporting UML profiles can be used.

The semantic analysis performed to obtain the corresponding equivalences allows identifying the classes of the activity diagram that are necessary to extend through stereotypes. The simulation software has a greater number of modules, besides those mentioned here, some of which are identified with other classes of UML. The elements that are part of the profile are a subset of the UML metamodel applied to a simulation environment. Therefore, the profile presented in this work extends the base classes: InitialNode, Action, DecisionNode, ActivityFinalNode, JoinNode, and ObjectNode ForkNode belonging to the activity diagram.

Although, every stereotype has its own properties, the user can edit the values of those properties. Figure 3shows a slice of the loading-transportation cycle, which highlights the properties of the stereotype ≪Create≫ and its corresponding values.

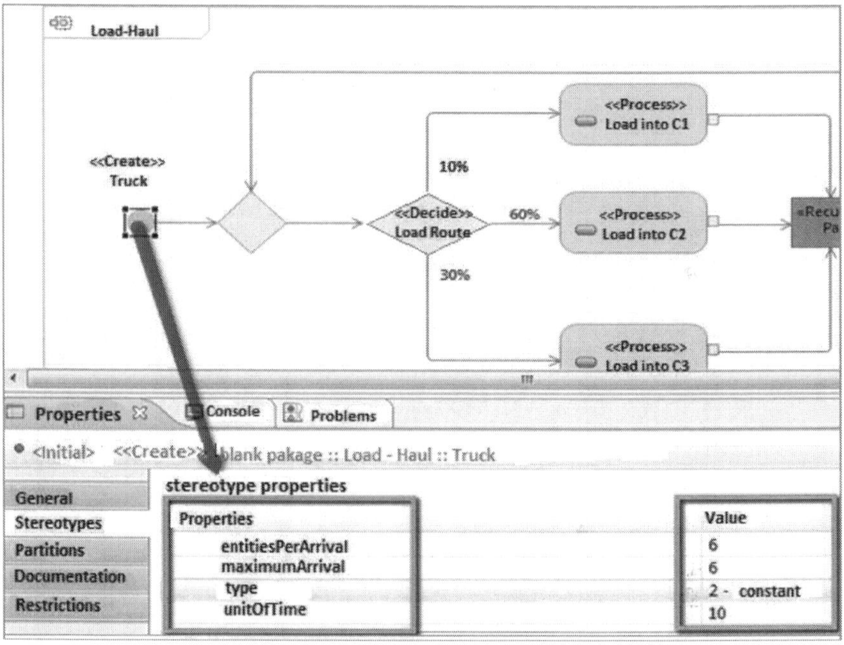

Figure 3. Properties of the stereotype ≪Create≫.

The stereotype ≪Create≫ extends the InitialNode metaclass. An InitialNode is a starting point for the implementation of an activity. In the UML specification, Superstructure v2.1.2 [5], the InitialNode is defined as a generalization of a control node where an object begins to flow through the system when an activity is invoked. This package belongs to the Basic Activities defined in the abstract syntax of UML Kernel Metamodel of OMG [5]. The Create module has attributes that define the input parameters, which must be defined by the user when starting the simulation. Figure 4 shows the stereotype and the extended metaclass. At the starting point of a simulation model we consider the (a) incoming flow rate generated (random or programmed constant) as well as the (b) number of entities that arrive to the system per unit time. In the mining process described in this work, every truck that enters into the system is a starting point.

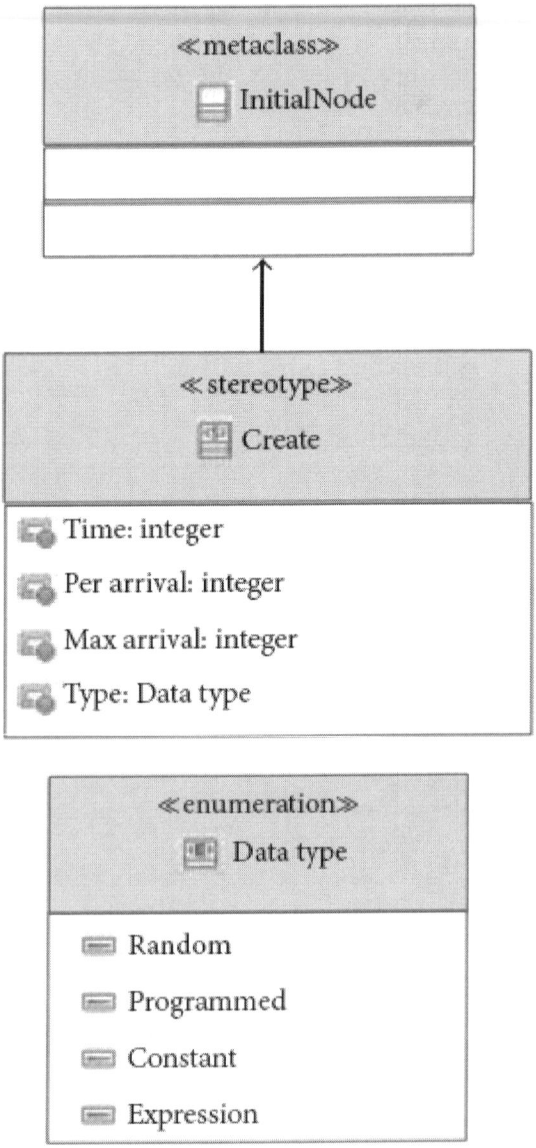

Figure 4. Relationship between metaclass and stereotype ≪Create≫.

Figure 5 shows the metamodel corresponding to the OMG specification. In this figure we show the metaclass named InitialNode which is a generalization of the metaclass named ControlNode.

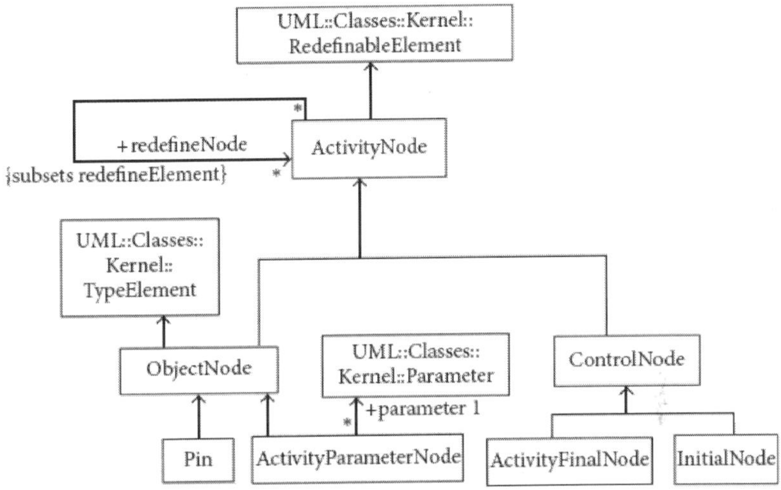

Figure 5. Metamodel of the OMG specification.

The stereotypes ≪Process≫, ≪Assign≫, and ≪Register≫ have a base class named Action [5] in the UML metamodel. For example, if we use an element of type ≪Process≫ in the UML model, it means that the stereotype ≪Process≫ behaves similarly to its base class. Additionally, it has to satisfy the semantics specified with OCL. Figure 6 shows the metaclass named ≪Action≫ and its corresponding stereotypes. For our particular case of study about mining processes related to load and transportation, the ≪Process≫ stereotype represents the load operation by means of the shovel.

The UML class named JoinNode [5] is extended through a so-called ≪Group≫ stereotype as shown in Figure7. In a simulation model, this class should specify the number of entities to be clustered as well as how these entities will be grouped (refers to whether to use random groups or, on the contrary, the entities should be grouped according to a specific attribute).

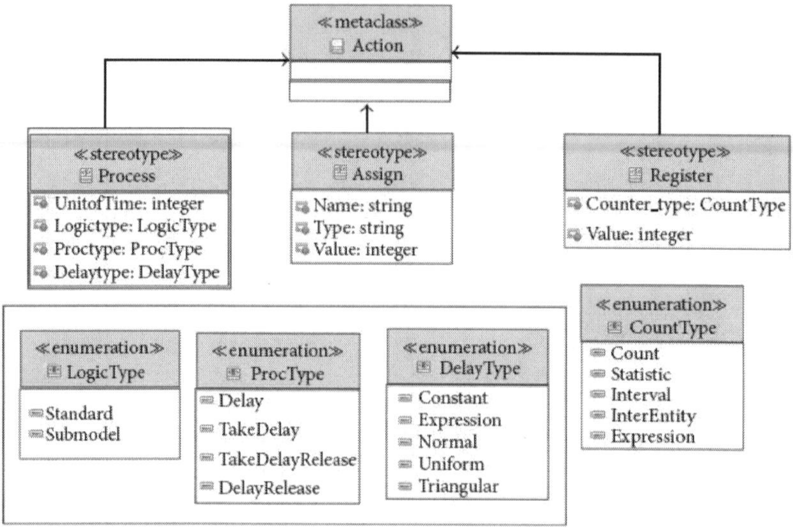

Figure 6. The metaclass action and its stereotypes.

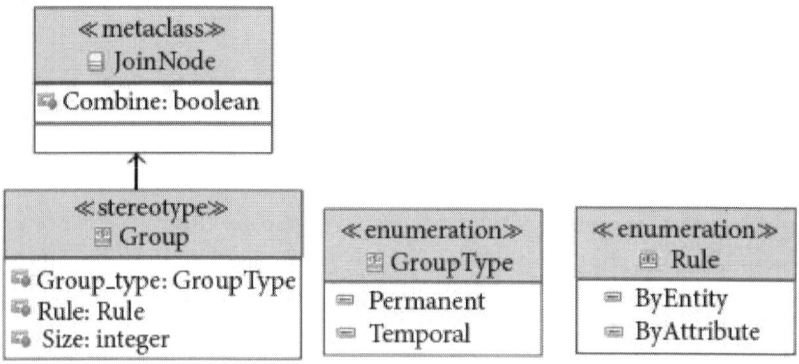

Figure 7. Metaclass JoinNode and its corresponding stereotype.

The class named ObjectNode is extended through four stereotypes ≪Entity≫, ≪WaitingQueue≫, ≪Resource≫, and ≪Set≫ as shown in Figure 8. The base class named ActivityFinalNode of the UML metamodel [5] was extended by a stereotype called ≪Dispose≫. The base class named ForkNode is extended through a stereotype called ≪Divide≫ [5]. This is shown in Figure 9.

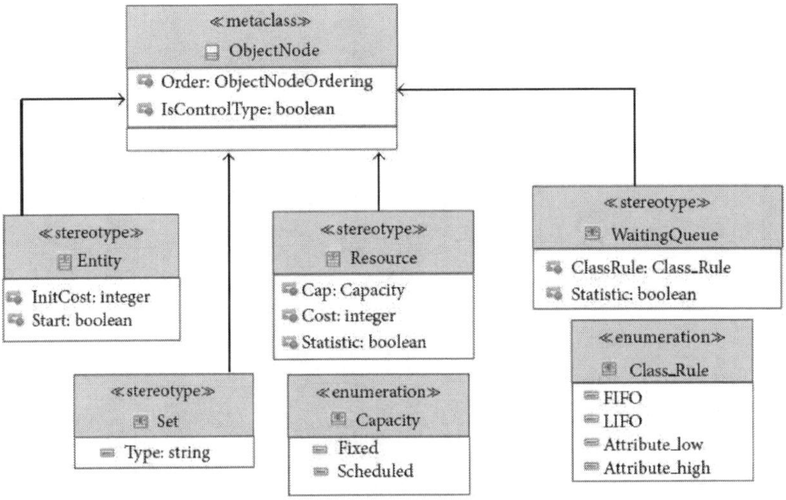

Figure 8. The ObjectNode metaclass with its stereotypes.

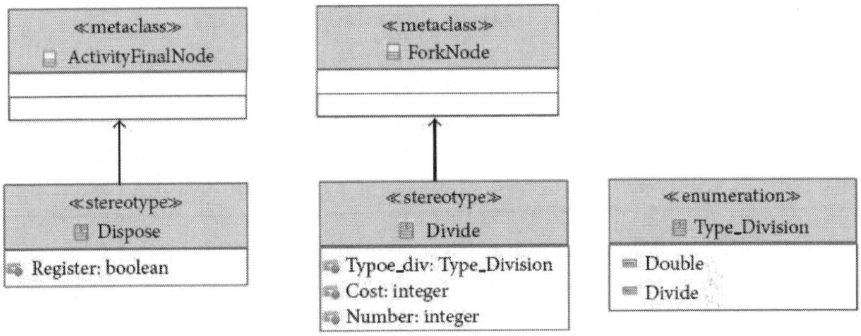

Figure 9. Metaclass ActivityFinalNode and ForkNode with its extended stereotypes.

The stereotype «Decide» [5] extends the UML metaclass named DecisionNode. Figure 10 shows this relationship.

The basic modules that integrate the simulation model of Arena are related to each other through connectors [10]. These connectors indicate how entities flow within the model. These connectors are represented with the metaclass of the activity diagram named ControlFlow. It is not necessary to extend this base class because there is a bilateral matching of the semantics.

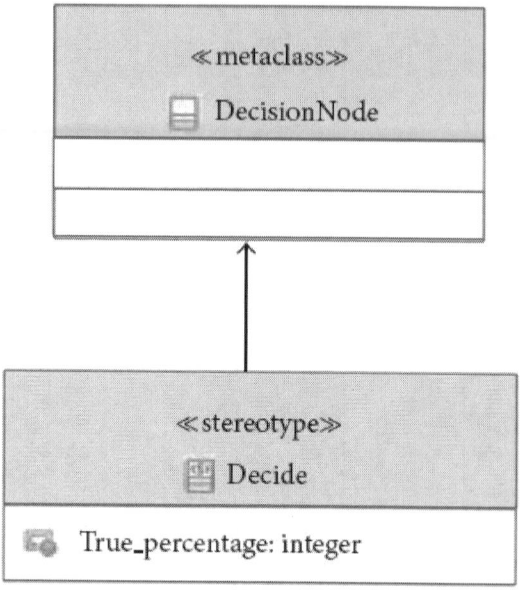

Figure 10. The metaclass DecisionNode and its stereotype.

Stereotypes defined above integrate the simulation profile that is applied in a mining process. The extended metaclasses were obtained as a result of observing the behavior of the basic modules of the simulation process. This analysis showed that while they share similarities in semantics, these modules have specific attributes that are not specified in the UML base classes, which is why we extended those classes, as described in the previous paragraphs.

5.1. OCL Restrictions

A model is an abstraction of a system. In particular in this work we model a mining process. It should be as accurate as possible. In order to find an unambiguous and consistent model, it is necessary to reformulate the restrictions to the developed profile.

The language proposed by UML to specify the restrictions of the diagrams is the so-called OCL (Object Constraint Language) [27]. OCL offers a convenient way of getting models with a high degree of consistency, because it is a formal language used to describe keywords in the model [34], making them more accurate and less ambiguous.

OCL is fully integrated into UML. We can check the values of the model elements and set restrictions on them. An example of a restriction on the profile element, in this case the stereotype ≪Decide≫, is especify in Algorithm 1.

Algorithm 1

The restrictions fulfill the purpose of validating the model where the proposed stereotypes apply. If an error occurs, it is immediately observed. In this case, we can introduce corrections so that the model complies with the preestablished conditions and reconcile with the expected results.

6. CONCLUSIONS

UML allows expressing specific concepts of discrete event simulation in the field of mining, through the extension of its basic classes. In particular, simulation software modules like CREATE and PROCESS are easily identified with the elements of the activity diagram. This feature provides high functionality to the proposed extended model, because its behavior is similar to the simulated system. System validation which was obtained by applying the profile to the mining process, where the load-transport cycle is a key part of a reservoir, satisfied the requirements imposed by the daily production.

Having a standard language like UML implies that it is possible to obtain a model with stereotypes with well-formed rules without ambiguous concepts. In this paper we presented an extension of UML for the basic elements of the Arena simulation software, which provides a higher level of abstraction. The integrated vision of these two languages (UML and simulation) helps build a consistent model for the process being simulated. Therefore, combining both tools facilitates the development of a system.

The proposed profile presents advantages at two levels.

Metamodel Level. At this level, any UML tool supporting profiles can be used to import the mining profile defined in this work. In this way, any engineer can use any UML tool for building their mining models using the same modeling language (UML profile). The second advantage is that if using a simulation language, other than Arena, or if the same Arena is extended with new characteristics, those characteristics can be easily included into the profile. The engineers use a modeling language (UML profile) independent of the simulation language (Arena, Simula, GPSS, etc.). The mining UML profile can be included with the specific semantic of any simulation language. Finally, using formal languages supporting mathematic specifications such as OCL (first order logic, theory of set and theory of bag) ensures much more accurate models than the ones used by other traditional modeling languages such as ERD (Entity Relationship Diagram), DFD (Data Flow Diagram) as general modeling language or

ExtendSim Suite, Arena Blocks as specific modeling languages. The mathematic specifications are incorporated as OCL specifications in the metamodeling level of the profile. Therefore, the proposed UML profile can be reused with any UML tool that includes profile definitions with OCL.

Model Level. At this level, the engineer builds its standard models using the proposed profile, and these models can be used in any mining environment. The stereotypes can be applied with any mining process simulation software (not only Arena). Moreover, any UML tool that allows incorporating the standard profiles according to the OMG can be used to modify these models. Finally, we obtained models verified mathematically with OCL. Those are more accurate models according to the restrictions that are specified in the profile level.

CONFLICT OF INTERESTS

The authors declare that there is no conflict of interests regarding the publication of this paper.

REFERENCES

1. OMG Unified Modeling Languaje (OMG UML), Infrastructure, Version2.3, OMG Document Number: formal/2010-05-03, http://www.omg.org/spec/UML/2.3/.
2. A. Watson, Visual Modeling: past, present and future, White paper UML Resource Page, 2008.
3. J. Miller and J. Mukerji, "MDA guide," Tech. Rep., Object Management Group, 2003, http://www.omg.org/mda. View at Google Scholar
4. G. Engels, J. H. Hausmann, R. Heckel, and S. Sauer, "Dynamic meta modeling: a graphical approach to the operational semantics of behavioural diagrams in UML," in Proceedings of the Unified Modeling Language (UML '00), vol. 1939 of Lecture Notes in Computer Science, pp. 323–337, 2000.
5. "OMG Unified Modeling Languaje (OMG UML)," Superestructure, Version2.1.2, OMG Document Number: formal/2007-11-02, Standard Document, 2007, http://www.omg.org/spec/UML/2.1.2/Superstructure/PDF.
6. J. Banks, J. S. Carson, and B. L. Nelson, Discrete-Event System Simulation, Prentice Hall, Upper Saddle River, NJ, USA, 2nd edition, 1996.

7. SimPy, Copyright 2002–2012 Team SimPy, http://simpy.readthedocs.org/en/latest/.

8. The Facsimile Simulation Library, Facsimile copyright 2004–2013, M. J. Allen, http://facsim.org/.

9. Elmer-CSC". CSC-IT Center for Science Ltd. Retrieved 2010-06-24.

10. W. D. Kelton, R. P. Sadoswski, and D. T. Sturrock, Simulación Con Software Arena, Cuarta Edición, McGraw-Hill, 2008.

11. M. N. Nashalji, M. A. Shoorehdeli, and M. Teshnehlab, "Fault detection of the Tennessee Eastman process using improved PCA and neural classifier," Advances in Intelligent and Soft Computing, vol. 75, pp. 41–50, 2010.

12. S. Yin, S. X. Ding, A. Haghani, H. Hao, and P. Zhang, "A comparison study of basic data-driven fault diagnosis and process monitoring methods on the benchmark Tennessee Eastman process," Journal of Process Control, vol. 22, no. 9, pp. 1567–1581, 2012.

13. S. Yin, H. Luo, and S. X. Ding, "Real-time implementation of fault-tolerant control systems with performance optimization," IEEE Transactions on Industrial Electronics, vol. 61, no. 5, pp. 2402–2411, 2013.

14. A. Teilans, A. Kleins, Y. Merkuryev, and A. Grinbergs, "Design of UML models and their simulation using Arena," WSEAS Transactions on Computer Research, vol. 3, no. 1, pp. 67–73, 2008.

15. J. Barjis and B. Shishkov, "UML based business systems modeling and simulation," in Proceedings of the 4th International EUROSIM Congress, June 2001.

16. O. Barbarisi and C. Del Vecchio, "UML simulation model for hybrid manufacturing systems," inProceedings of the13th Mediterranean Conference on Control and Automation, Limassol, Cyprus, June 2005.

17. H. J. Pels and J. Goossenaerts, "A conceptual modeling technique for discrete event simulation of operational processes," IFIP International Federation for Information Processing, vol. 246, pp. 305–312, 2007.

18. L. Fuentes-Fernández and A. Vallecillo-Moreno, "An introduction to UML profiles," The European Journal for the Informatics Professional, vol. 2, no. 2, pp. 6–13, 2004.

19. Object Management Group, "Common warehouse metamodel (CWM) specification," OMG Documentad/2001-02-01, 2001.

20. N. C. Debnath, A. Garis, D. Riesco, and G. Montejano, "Defining patterns using UML profiles," inProceedings of the IEEE

International Conference on Computer Systems and Applications, pp. 1147–1150, March 2006.

21. N. Debnath, C. A. Martinez, F. Zorzan, D. Riesco, and G. Montejano, "Transformation of business process models BPMN 2.0 into components of the Java business platform," in Proceedings of the 10th IEEE International Conference on, Industrial Informatics (INDIN), pp. 1035–1040, July 2012.

22. N. C. Debnath, L. Baigorria, D. Riesco, and G. Montejano, "Metrics applied to aspect oriented design using UML profiles," in Proceedings of the 13th IEEE Symposium on Computers and Communications (ISCC '08), pp. 654–657, July 2008.

23. Matlab, 1994–2013 The MathWorks, Inc, http://www.mathworks.com/products/matlab/.

24. Y. Merkuryev, Riga Technical University. The Modelling and Simulation of Complex Systems: Methodology and Practice. An Overview. Information Technology and Management Science, 2012.

25. R. Castro, E. Kofman, and G. Wainer, "A formal framework for stochastic discrete event system specification modeling and simulation," Simulation, vol. 86, no. 10, pp. 587–611, 2010.

26. A. Kleins, Y. Merkuryev, A. Teilans, and M. Filonix, "A meta-model based approach to UML modelling and simulation," in Proceedings of the 7th WSEAS International Conference on System Science and Simulation in Engineering (ICOSSSE '08), 2008.

27. OMG Object Constraint Languaje (OCL), OMG Document Number: formal/2012-01-01. Standard document, http://www.omg.org/spec/OCL/2.3.1.

28. M. Bustillo Revuelta and C. López Jimeno, Recursos Minerales. Tipología, Prospección, Evaluación, Explotación, Mineralurgia, Impacto Ambiental. Madrid, 1996.

29. M. Bustillo Revuelta and C. López Jimeno, Manual de Evaluación y Diseño de Explotaciones Mineras. Madrid, 1997.

30. M. S. Abdullah, R. Paige, C. Kimble, and I. Benest, "A UML profile for knowledge-based systems modelling," in Proceedings of the IEEE 5th ACIS International Conference on Software Engineering Research, Management, and Applications (SERA '07), pp. 871–878, August 2007.

31. C. Bartolini, A. Bertolino, G. De Angelis, and G. Lipari, "A UML profile and a methodology for real-time systems design," in Proceedings of the IEEE 32nd Euromicro Conference on Software Engineering and Advanced Applications (SEAA '06), pp. 108–115, September 2006.

32. D. S. Frankel, Model Driven Arhcitecture. Applying MDA to Enterprise Computing, 2003.

33. IBM Rational Software Architect version 8.5. Copyright IBM Corporation 1987, 2012, http://www-01.ibm.com/software/rational/.

34. UML 2.0 OCL Specification, 2006, http://www.omg.org.

7

MODELLING UNDERGROUND COAL GASIFICATION—A REVIEW

Md M. Khan [1], Joseph P. Mmbaga [1], Ahad S. Shirazi [1], Japan Trivedi [2], Qingxia Liu [1] and Rajender Gupta [1]

[1]Department of Chemical and Materials Engineering, University of Alberta, Donadeo Innovation Centre for Engineering, 9211–116 Street NW, Edmonton, AB T6G 1H9, Canada

[2]School of Mining and Petroleum Engineering, University of Alberta, Edmonton, AB T6G 1H9, Canada

ABSTRACT

The technical feasibility of underground coal gasification (UCG) has been established through many field trials and laboratory-scale experiments over the past decades. However, the UCG is site specific and the commercialization of UCG is being hindered due to the lack of complete information for a specific site of operation. Since conducting UCG trials and data extraction are costly and difficult, modeling has been an important part of UCG study to predict the effect of various physical and operating parameters on the performance of the process. Over the years, various models have been developed in order to improve the understanding of the UCG process. This article reviews the approaches, key concepts, assumptions, and limitations of various forward gasification UCG models for cavity growth and product gas recovery. However, emphasis is given to the most important models, such as packed bed models, the channel model, and the coal slab model. In addition, because of the integral part of the main models, various sub-models such as drying and pyrolysis are also included in this review. The aim of this study is to provide an overview of the various simulation methodologies and sub-models in order to enhance the understanding of the critical aspects of the UCG process.

Keywords. underground coal gasification; UCG modeling; coal

1. INTRODUCTION

Coal is the most abundant and widespread type of fossil fuel, with reserves of economically recoverable coal estimated to be about 900 gigatons (GT), 64% of major fossil fuels [1]. However, the total amount of coal potential including the economically non-recoverable part is reported to be 18 terratons [2], most of which are too deep or too costly to be mined by traditional methods. With the current trends of energy consumption, and the rise of developing countries, the total fossil fuel reserve is expected to be the same while the total energy consumption is increasing around the world. In order to increase the recoverable fossil fuel reserve, it is important to find appropriate technology to extract the non-recoverable part of the reserves economically. Underground coal gasification is a method of *in-situ* conversion of deep un-minable coal into gaseous products with a higher heating value through controlled combustion with oxygen/air and gasification with steam. The UCG product gas, called synthesis gas (syngas), is mainly a mixture of CO, H_2, CH_4, and CO_2 with trace amounts of ethane and other components. The development of the underground (*in-situ*) coal gasification (UCG) technique is paving the way to access deep coal seams. Early studies suggest that the use of UCG could potentially increase the world's coal reserves by as much as 600 GT [3], which represents a 70% increase [2].

1.1. Environmental and Safety Features of UCG and Coal Mining

The technology of coal mining and the use of coal have been affected by the pollution caused by its transport, storage, and combustion [4]. In addition, coal-fired power plants have higher greenhouse gas (GHG) emission per generated energy due to lower efficiency of the process and a higher carbon-to-hydrogen ratio of coal. However, the most alarming issue is the worldwide death fatalities and injuries per year in the coal mining industries. It has been reported that there was a total 36,910 deaths in a seven-year period, from 2000 to 2005, in the coal mining industries of China alone [5]. On the other hand, the USA has reduced their death fatalities after ensuring safe work environments in the current century; however, the death fatalities of the USA have still numbered more than 300 since 2003 [6]. Considering the death fatalities alone, UCG is the safest energy harness process. Moreover, UCG has several advantages over the conventional gasification process. Examples of such benefits include [7,8,9]:

- Elimination of coal mining and thus avoiding a dangerous practice even with current equipment and safety regulations.
- Lower capital investment due to the absence of surface gasification units.
- Elimination of ash handling operations by keeping the ash underground.
- Lower water consumption.
- Generation of possible sites for CO_2 sequestration.

From the economic and environmental point of view for large-scale gas production, UCG is a highly promising technology [7,10]; however, UCG introduces some challenges that should be addressed before the process can be adopted on a large scale. Some of the main challenges are process stability, aquifer contamination, and subsidence [11]. However, the potential risks of subsidence and aquifer contamination can be avoided through careful site selection and adoption of best management practices for operations [7,8,11,12].

1.2. UCG Process Overview

The UCG process involves four main steps, namely drilling, linking, ignition, and gasification. The basic layout of the process requires one borehole for the injection of gases and one for production; however, over the years three standard configurations of wells have evolved. They are:

- Linked vertical wells (LVW);
- Controlled retraction injection point (CRIP);
- Steeply dipping seams (SDS).

Figure 1 shows a sketch of linked vertical wells arrangements for the UCG process. In the first step, injection and production wells apart from each other are drilled at some specified distance from the surface into the coal seam. In the following step, a permeable channel (link) is created between two wells which allows the sufficient flow of gases in a large-scale operation. It is necessary to ignite the coal initially in order to raise the coal temperature using combustible gas and oxygen mixture with the help of a suitable ignition source. The duration of this combustion stage depends on the seam properties as well as the operating conditions. After this stage, the mixture of air/oxygen and steam are injected into the seam to gasify the coal. The main product of gasification, syngas, is then recovered from the production well. As the coal is being consumed, ash falls down and a void space is created, which is referred to as a cavity. When the cavity reaches the production well, there will be less contact between the injection gases and the virgin coal; thus, the quality of produced gases drops significantly. Therefore, fresh coal should be made available by moving the location of the injection gases for further continuation of the process. After the completion of gasification, steam and/or water are usually flushed to the cavity in order to remove the pollutants from the coal seam to prevent them from diffusing into surrounding aquifers.

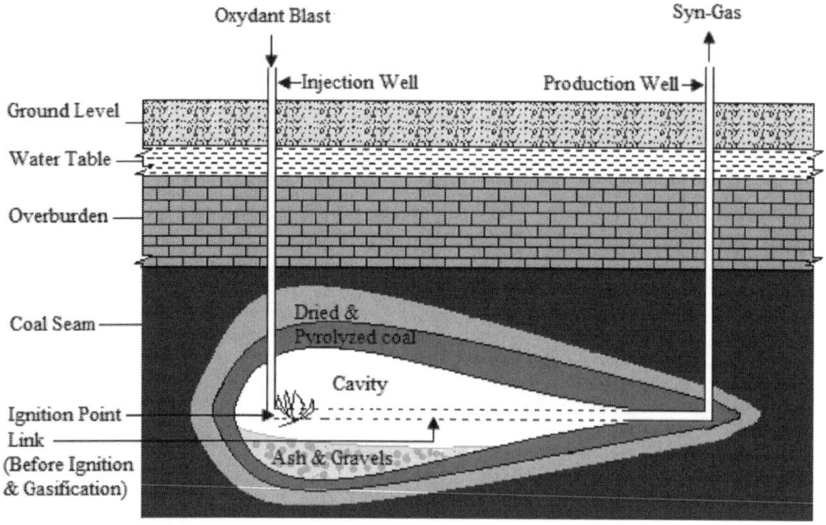

Figure 1. Schematic diagram (not to scale) of the UCG process for linked vertical wells (LVW).

1.3. Chemical Reactions in UCG

The UCG process occurs in three different zones: drying, pyrolysis, and combustion and gasification of solid char. In the drying zone, wet coal is heated and converted into dry coal by removing moisture attached to it. Upon further heating of the coal, the pyrolysis reactions begin at temperatures around 350–400 °C [13] and coal loses its weight, generating volatile matters and the solid known as char. Finally, the char reacts with the injected/pyrolyzed gases to produce the syngas. The three essential types of char layer reactions which govern the UCG product gas composition are: oxidation, reduction, and pyrolysis. Many reactions occur during this process, but the most important reactions, which are considered in most of the models, are summarized in Table 1.

Table 1. Main reactions in the gasification process.

Reaction	Reaction No.	Representation	Heat of Reaction (kJ/mol)
Drying	1	Coal → Dry Coal+ H_2O	+40
Pyrolysis	2	Dry Coal → Char + Volatiles	0
Coal combustion	3	C (Char) + O_2 → CO_2	−393
CO_2 gasification	4	C (Char) + CO_2 → 2CO	+172
Steam gasification	5	C(Char) + H_2O → CO+ H_2	+131
Methanation	6	C(Char) + $2H_2$ → CH_4	−75
Water-gas shift	7	CO + H_2O ↔ CO_2+ H_2	−41
	8	CO + $0.5O_2$ → CO_2	−111
Gas phase oxidations	9	H_2 + $0.5O_2$ → H_2O	−242
	10	CH_4 + $2O_2$→ CO_2 + $2H_2O$	−802

The reactivity of the char to O_2, H_2O, CO_2, and H_2 determines the rates at which the desired products are formed [9]. Reactions 3, 5, and 8 are the main chemical reactions considered for both shallow and deep coal gasification processes. The hydrogasification (reaction 6) is favourable at a high hydrogen pressure. In the UCG, at low pressure, this reaction is not significant. Reactions 3–6 take place on the wall of coal seams, while reactions 7 and 8 occur at the gaseous stage [14]. Different reactions are favourable at different temperature ranges. Figure 2 shows the equilibrium constants as a function of temperature for some of the important reactions in the gasification process. As the equilibrium constants for all the reactions shown in Figure 2 are at unity at approximately 1000 K, the temperature becomes the key factor in determining the shift of chemical equilibria towards the product or reactions by deciding whether it is an endothermic or exothermic reaction. For example, at temperatures above 1000 K, the formations of H_2 and CO are favourable since reaction 4 and reaction 5 are endothermic [15]. Similarly, other reactions seem to be favoured at temperatures below 1000 K. Reactions involved in UCG are not fundamentally different from the reactions of coal and char in the surface gasifiers. In gasification reactions, the combustion of coal and volatiles provides the necessary heat for endothermic gasification reactions.

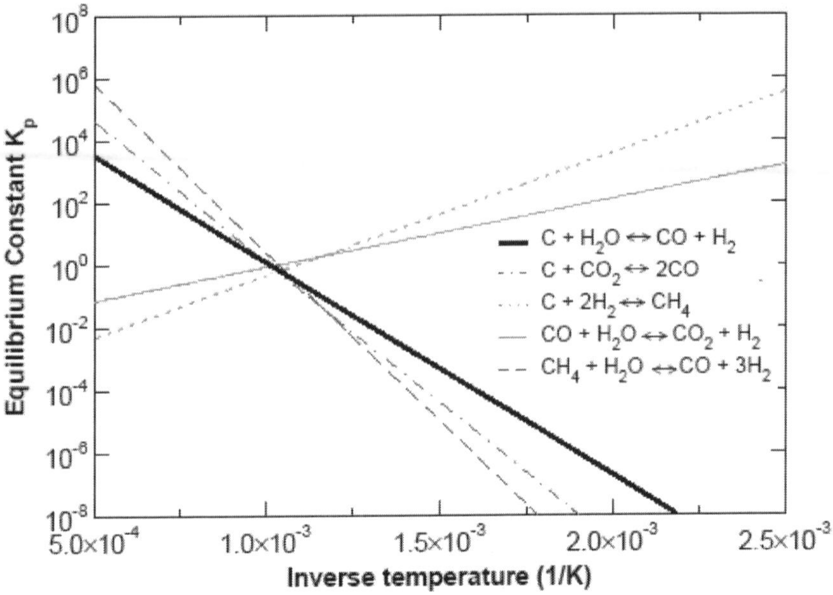

Figure 2. Equilibrium constant *Kp* as a function of temperature for some reactions in coal gasification [15].

1.4. Brief History of UCG Practice

Although the idea of the UCG was first mentioned by Sir William Siemens of Great Britain in 1868 [16,17,18] and the idea was consolidated (1988–1899) independently by the famous Russian chemist Dmitri I. Mendeleev, who also pointed out the economic benefit of UCG over conventional mining [16], most credit goes to an American chemist, A.G. Betts, due to his detailed technical contents and engineering drawing on the UCG method that closely resemble modern approaches presented in his three UCG patents received during 1909–1910 [16,17]. Later on, in 1912, famous British chemist Sir William Ramsay expanded the ideas of Betts and proposed gasifying coal underground as a way to avoid air pollution and as a solution for the mining worker [16]. His strong advocacy for the development of this technology created an international surge of interest in UCG that eventually helped him to culminate the first ever UCG experiment being carried out at Hett Hill near Tursdale Colliery in County Durham, North East England, in 1912 [16,18]. Unfortunately, the experiments did not proceed due to the outbreak of World War I in 1914 and Ramsay's death in 1916. A detailed description of early ideas in underground coal gasification and their evolution is given in the excellent historic review of UCG by Klimenko [16].

After successful operations of UCG throughout the 1930s and 1960s, the process was left out mostly because of the low prices of gas and oil. Due to increased energy demands and stringent environmental regulations, UCG has regained its popularity in recent years as an immediate alternative for conventional coal gasification plants. Over 50 pilot UCG plants have been conducted worldwide since the 1930s. Most of these developments have been concentrated in the former USSR, Europe, the USA, and China. The former USSR was the first country to operate UCG successfully and they gained extensive knowledge in the UCG process. However, their knowledge was scarcely available until the 1960s. Most of the information on Russian tests was available in the Russian literature. As a result, most of the field trials conducted by other countries before the 1960s, without the benefit of detailed data on Soviet field tests, did little to advance the process and left an impression that UCG was not a viable process [19]. Later on, many Russian articles on UCG were translated into English. As a result, the volumes of works on UCG carried out by the former USSR were exposed to the rest of the world. After getting a complete picture of the Soviet UCG experience with the large-scale translation of Russian technical literature, Gregg *et al.* [19] concluded that "the amount of effort expended and degree of success achieved by the Soviets (having constructed and operated several commercial plants) far exceeds the summation of efforts by other nations". Translations of a Russian book and summaries of the translations of Russian articles and specific experiences on UCG can be found in several references: Skafa [20], Gregg *et al.* [21], Derbin *et al.* [22]. After gaining knowledge from the Soviets' field practices, extensive field trials were performed in the USA and large-scale pilots proved the basic feasibility of UCG [23]. Earlier, the field tests were limited to shallow depth, following the Soviet practice, but over time, with the improvements in drilling technology, the depth of UCG operations has increased markedly as can be observed in recent European field trials [11]. The underground preparation of the seam and gasification are performed using drilled holes. Earlier, linked vertical wells (LVW) and steeply dipping seams (SDS) were the most popular drilling configurations for horizontal and steeply dipping seams, respectively. However, recently, the controlled retraction injection point (CRIP) invented in the Centralia series of UCG trials in the 1980s has gained considerable attraction due to its greater control and improved overall efficiency [2,10]. Establishing a highly permeable path between the injection and production holes is considered as the single most important task prior to gasification because a highly permeable narrow channel can control gas leakage, water intrusion, liquid condensation, phenol contamination, and oxidant bypass during UCG operation [19]. Directional drilling, reverse combustion, and hydrofracturing are the popular linking methods currently [2,19]. After the invention of reverse combustion linking (RCL) in 1941 by former Soviet researchers [24], the technique was widely used by many countries in UCG field trials, including the former USSR. Reverse combustion linking (RCL) is the opposite of forward combustion linking (FCL) where the coal is ignited at the base of the injection well and the flame propagates towards the production well, *i.e.*, in the direction opposite to the gas flow within the channel [24,25]. As compared to FCL, RCL is the preferable linking option in terms of channel dimension, consumption of coal per unit of the link, linking speed, operational stability, and controllability [24]. However, FCL can also be used for widening the hydraulic links between the wells which exist naturally or have been preliminarily established using, for example, hydrofracturing [24]. Hydrofracturing is employed to establish a linking channel in coal seam of low permeability where RCL alone is not a suitable option initially due to *in-situ* coal properties and the fact that directional drilling is too sophisticated for that situation. The effectiveness of hydrofracturing is independent of both the electrical properties and the natural permeability of the coal; however, its lack of control limits its use primarily for steeply dipping beds where it is not essential to ensure that the cracks form at the bottom of the seam or that they form parallel or perpendicular to the bedding planes [19]. Directional drilling is one technique that is relatively independent of the coal properties. The in-seam drilling of coal seams has been part of coal exploitation since at least the 1940s in Soviet plants [19]. Due to continuous and significant advances in the techniques used for directional drilling, this method is being currently considered as the most reliable way of establishing the link between wells.

UCG field tests have been carried in different coal seams worldwide, using different techniques in order to evaluate the effects of various factors of overall UCG performance. Some significant field trials worldwide are summarised in Table 2. A quick look at Table 2 indicates a significant increased heating value of the product gas for an injection gas of oxygen/steam mixture over air. Moreover, the overall performance is largely affected by coal properties (*i.e.*, ranks and chemical composition), coal seam geology (*i.e.*, depth, thickness), and drilling configurations. The dependence of coal properties, coal seam geology, and overburden properties (geology and hydrology) indicates that the performance of UCG is site-specific. The site selection is perhaps the single most important decision that will determine the technical and economic feasibility of coal gasification [15]. However, for same site, seam preparation and operating conditions (*i.e.*, temperature, pressure) are the key factors for the successful performance of UCG.

Table 2. Summary of significant underground coal gasification (UCG) field trials worldwide [2,15,26,27,28,29].

Country	Site	Start-Up Year	Coal Rank	Well Config-uration	Injected Gas	Seam Depth (m)	Seam Thickness (m)	Gas Calorific Value (MJ/m^3)
USSR (former)	Gorlovka (Ukraine)	1937	B	SDS	Air	40	1.9	3.95
					O_2/H_2O	40	1.9	5.82–10.34
	Podmoskova/Tula (Russia)	1947	L	LVW	Air	55	3	3.4
	Yuzhno-Abinsk	1955	B	LVW	Air	~100	2	4.1
	Angren (Uzbekistan)	1961	L	LVW	Air	110	4	3.5–3.9
	Hanna II	1975	B	LVW	Air	85	9	4.2
USA	Hoe Creek I	1976	B	LVW	Air	40	8	3.6
	Hoe Creek IIB	1977	B	LVW	O_2/H_2O	40	8	9.0
	Price town 1	1979	B	LVW	Air	270	2	6.1
	Rawlins IA	1979	SB	SDS	Air	105	18	5.6
	Rawlins IIB	1979	SB	SDS	O_2/H_2O	105	18	8.1
	Centralia A	1984	SB	CRIP	O_2/H_2O	75	6	9.7
	Centralia B	1984	SB	LVW	O_2/H_2O	75	6	8.4
	Rocky Mountain IA	1987	SB	CRIP	O_2/H_2O	110	7	9.5
	Rocky Mountain IB	1987	SB	LVW	O_2/H_2O	110	7	8.8
	Alaska SHR	2012	L/SB		Air	50–650	1–12	3.3–5
Belgium	Thulin	1986	A	LVW	Air	860	6	7.0
Spain	El Tremedal	1997	SB	CRIP	O_2/H_2O	580	2	6.6
Australia	Chinchilla	2000	SB	LVW	Air	140	10	6.6
	Bloodwood	2011		CRIP	Air	200	8-10	6.4
China	Suncan	2002	B	Tunnel	Air	80	2	8.5
South A.	Majuba	2007	B	LVW	Air	285	3.5–4.5	6.2

B = Bituminous, SB = Subbituminous, A = Anthracite, L = Lignite.

2. UCG MODELING

Although a number of UCG field trials have been performed, the information on the detailed UCG process is very limited because of the high cost of extracting data as well as the difficulty in controlling the operating variables. In addition, there is a limited scope of adjusting parameters due to the site-specific nature of the performance of UCG. As a result, due to relative economic consideration, several laboratory-scale coal block experiments have been reported [30,31]. Table 3 represents major laboratory studies on UCG. Feed temperature, feed flow rate, steam-to-oxygen ratio, operating pressure, combustion time, total operation time, and distance between wells are varied in order to observe their effects on the product gas composition and flow rate and gas outlet temperature. As well, the temperature distribution in the coal seam during gasification, the final cavity dimension, and the rate of recession of the coal surface due to reaction are also observed and, finally, the optimum operating conditions for the expected UCG performance are determined.

Table 3. Major UCG experiments on coal blocks [27].

Researchers	Coal Type	Dimensions of Seam (Length × Width × Height)	Major Observations
Yeary and Riggs [32]	Lignite- Sub-bituminous	25 cm × 5 cm × 25 cm	Lateral cavity growth
Poon [33]	Lignite	9 cm × 2.5 cm	Lateral cavity growth
Daggupati *et al.* [34]	Lignite	30-38 cm × 20cm × 25cm	Cavity growth and shape
Yang [35]	High-volatile bituminous	6.8 m × 0.25 m × 1.1 m (68°) 7.5 m × 2 m × 1 m (65°)	Two-phase UCG in inclined abandoned coal mines
Prabu and Jaynati [36]	Camphor, Wood, Lignite	24 cm × 8–10 cm × 10–12 cm	Cavity shape
Stanczyk *et al.* [37]	Lignite- Hard Coal	2.5m × 0.7m × 0.7m	Two-phase UCG in lignite

The generalizing data from the lab-scale experiments should still be done with extreme care since the whole process may not be properly represented in lab-scale experiments. It may not be possible to represent the entire phenomenon in laboratory-scale experiments [38,39]. The coal seam is a three-dimensional reactor without specific geometry and the flow pattern inside the UCG cavity is highly non-ideal. The complexity increases further because of several other complex physical and chemical processes occurring simultaneously, such as chemical reactions and kinetics, transport phenomena (*i.e.*, heat and mass), water intrusion from surrounding strata, thermo-mechanical processes related to the structure of the seam, and other geological aspects. Both field tests and lab-scale experiments are not sufficient to provide a detailed quantitative description due to the demand of extensive instrumentation and trials. The difficulties encountered in controlling and operating the UCG process serve to be the incentive to develop quantitative models in order to predict the effects of various physical and operating parameters on the performance of the process and to extrapolate the understanding to other operating conditions and coals. Prior to 1975, the development of UCG models was very limited [40], and focused only isolated cases, such as heat transfer, mass transfer, combustion rate, *etc.* However, after gaining a better understanding of the UCG process, the intensity of developing a global model increased significantly. A full UCG model that describes the complete mechanisms would typically include a number of the following sub-models:

- Injection/production linkage sub-model.
- UCG reactor sub-model.
- Ground water hydrology.
- Ground subsidence model.
- Surface facility models.

Combining all these sub-models would theoretically give an exact description of the process; however, building such a model is not a trivial task. Therefore, all previous models have focused on studying these aspects separately through several simplifying assumptions. There are lots of models developed thus far; however, a successful model should be able to:

1. Compute transient temperature profiles along the coal seam during gasification.
2. Determine the rate of gas and coal consumption.
3. Determine the effect of coal shrinkage or swelling on UCG performance.

4. Predict the pressure and velocity of the produced gas in a coal seam of known porosity and permeability.
5. Predict the progressive configuration of the combustion front during gasification.
6. Predict the cavity formation and dimension with time.
7. Predict the effect of operating pressure, feed inlet temperature, feed rate, feed mixture ratio, and well spacing and configuration for the gas production rate, composition, and its heating value as well as cavity dimension.

1. Predict the effect of coal bed thickness, ash content, moisture content for the gas production rate, composition, and its heating value as well as cavity dimension.

In this article, a brief overview of the UCG reactor sub-model of forward gasification is particularly discussed due to its direct impacts on coal resource recovery, cavity growth, energy efficiency and, consequently, economic feasibility.

Over the years, several approaches have been developed for the modeling of the UCG process; however, the main approaches are:

- Packed bed model.
- Channel model.
- Coal slab model.

Since there are a number of mathematical/numerical models in the literature, it is most instructive to indicate the significant features and limitations of these models. Most of the earlier models were one-dimensional (1D); however, with the advancement of computational power, two-dimensional (2D) and even a few three-dimensional (3D) models were developed. In the following section, the various modeling approaches are discussed.

2.1. Packed Bed Models

The earliest models of UCG in literature include models that describe the process as a packed bed reactor. The consideration of the "*in-situ*" coal seam as a packed bed primarily originated from the concept of Higgins [41] considering the creation of a permeable zone between two boreholes either by reverse combustion linking (RCL) or by fracturing the coal seam using pressurized air or chemical explosives [42,43]. The resultant coal seam resembles a packed bed where coal particles are filled in the reactor. This concept is similar to the major Soviet approach to seam preparation where they included extensive drying of the seam and reverse combustion to obtain a region of enhanced permeability between the boreholes [44]. On the other hand, according to Gunn and Whitman [45], the coal seam consists of lignite or subbituminous coal can be gasified without establishing any linkage path due to their high permeability, also known as the percolation or filtration method which more closely resembles the operation of a packed bed chemical reactor. The packed bed model assumes that coal gasification occurs in highly permeable porous media with a stationary coal bed which is consumed over time [46]. Table 4 and Table 5 represent a glimpse of a list of packed bed models discussed here with their essential features and reaction rate control mechanisms of the main reactions in the gasification process, respectively.

Initially, around 1976, several 1D transient packed bed models [42,47,48] with the finite-difference approach were developed in parallel, supported by the Lawrence Livermore National Laboratory (LLNL) program. However, there is no basic difference between the models developed by Thorsness *et al.* [48] and Thorsness and Rozsa [42]. Thorsness and Rozsa [42] provided a detailed description of the lab-scale gasification experiment whereas Thorsness*et al.* [48] provided a detailed description of the development of the physical and chemical model where they made many simplifying physical and chemical assumptions, based on experimental and theoretical comparisons and correlations, which include:

- the gas permeability.
- the effective thermal conductivity.
- the interphase heat-transfer coefficient.
- the chemical reaction rates.
- the various thermodynamic properties of each species, and.
- the stoichiometric coefficients.

In their model, they neglected tar condensation and plugging, gas losses to surroundings, water intrusion from surroundings, heat losses, and coal bed movement due to shrinking or swelling during drying and pyrolysis in order to avoid complexity.

Winslow [47] also followed their work, except using a different numerical solution scheme. Thorsness *et al.* [48] considered all the main reactions for gasification (reaction 1 to reaction 10); however, for homogeneous reactions, only the water-gas shift reaction was considered by Winslow [47]. Although, for convenience, most of the researchers represent char as a molecule containing one carbon, they considered the chemical formula of char and dry coal as $CH_{a1}O_{b1}$ and $CH_{a2}O_{b2}$, respectively, where the composition parameters a1, b1, a2, b2 depend on the type of coal (usually obtained from the ultimate analysis). Beside the moisture content for drying, they also considered mobile condensed water which undergoes evaporation and condensation upon the removal of heating. However, for the simplicity of the calculation, they assumed a fixed velocity of mobile water. Their model considered fitted kinetic models for pyrolysis data developed by Campbell [49]. The controlling mechanisms of reaction rates that they considered are listed in Table 5. Eight gas species (N_2, O_2, H_2O, H_2, CH_4, CO, CO_2, tar) and five solid/condensed species (coal, char, mobile liquid water, fixed liquid water, and ash) were considered, upon which the balances were written. The reaction rates of the heterogeneous reactions (char-gas reactions) were considered to be influenced by three mechanisms: mass-transport limitation from the bulk gas to the solid surface, mass-transport limitations in any internal particle porosity, and kinetic limitations at the solid surface. This reaction mechanism was also considered for the water-gas shift reaction due to the influence of a catalyst that might be present in the coal bed. Because of the high reactivity of char as compared to pure carbon, the mass transfer limitation was considered important. However, in their model, one effective chemical rate expression (R_c) was assumed in order to account for the internal transport and surface kinetics. As a result, the total rate (R_T) was considered as a kinetic internal-diffusion rate and expressed as follows:

$$\frac{1}{R_T} = \frac{1}{R_c} + \frac{1}{R_m} \tag{1}$$

where R_c is linearly dependant on the mole fraction of the limiting reactant (y_l) in the bulk phase and the parameter R_m is the mass transfer rate of a limiting reaction defined by:

$$R_m = k_y y_l \tag{2}$$

where k_y is the mass transfer coefficient.

In their models, the fluid flow was described by the Darcy equation, in which permeability was set to change with the extent of the reactions. From the experimental data of drying and pyrolysis of subbituminous Wyodak coal, they formulated a functional relationship relating permeability to porosity that was used later by a number of researchers

[12,47,50] who developed their model using subbituminous coal. The relationship is as follows:

$$\alpha = \alpha_0 \exp\left[\sigma\left(\varphi - \varphi_0\right)\right] \text{ for} \varphi > \varphi_{\text{lim}} \tag{3}$$

where α_o is the initial permeability, φ_o is initial porosity of coal, and φ_{lim} is the upper limit of porosity above which the permeability was assumed to be constant. Based on the experiment, the calculated value of the parameter σ was approximately 12. In their report, they considered $\varphi_{lim} = 0.25$ so that the particle size decreases with increasing porosity ($\varphi < 0.25$). As a result, the permeability was assumed to reach a constant value for porosity larger than 0.25.

Thorsness *et al.* [48] incorporated a quasi-steady approximation for changes associated with the gas phase that allowed them to use ordinary differential equations for gas phase mass and energy balances. The governing equations were as follows:

Mass balance:

$$\text{Gas: } \frac{duC_i}{dx} = \Sigma_j a_{ij} R_j \; (i = 1, 2, \ldots . G) \tag{4}$$

where G is the number of gas species.

It is noted that gas phase diffusion, which had been considered an important factor in the channel and coal slab model development, was neglected.

$$\text{Solid: } \frac{\partial \rho_k}{\partial t} = M_k \Sigma_j a_{kj} R_j \; (k = 1, 2, 3 \ldots . K) \tag{5}$$

where K is the number of solid/condensed species.

$$\text{Mobile water: } \frac{\partial \rho_w}{\partial t} = -u_w \frac{\partial \rho_w}{\partial x} + M_w \Sigma_j a_{kj} R_j \; (k = 1, \textit{ i.e., only water}) \tag{6}$$

Energy balance:

$$\text{Gas: } \frac{dT}{dx} = -\frac{1}{uC_{p_g}} \left[h_T(T - T_s] + Q_g\right. \tag{7}$$

$$\text{Solid: } C_{p_s} \frac{\partial T_s}{\partial t} = \frac{\partial}{\partial x}\left[(1-\varphi) D \frac{\partial T_s}{\partial x}\right] + h_T\left(T - T_s\right) + Q_s - u_w \rho_w C_w . \frac{\partial T_s}{\partial x} \tag{8}$$

Momentum balance:

The momentum balance gas phase was replaced by Darcy's law:

$$\frac{\partial p}{\partial x} = -\frac{u}{\alpha}\mu_g \tag{9}$$

The use of Darcy's law and some theoretical correlations they employed in their model indicate an implicit assumption of the laminar flow process.

On the other hand, Winslow [47] did not approximate any stationary phase for the basic conservation equation, although he followed the physical and chemical model of Thorsness *et al.* [48]. The following governing equations for mass and energy balance for both the gas and solid phases were used by Winslow [47]:

$$\text{Overall gas conservation: } \frac{\partial}{\partial t}(\varphi \rho_g) = -\frac{\partial}{\partial x}(\rho_g u) + S_m \tag{10}$$

$$\text{Conservation of gas species: } \frac{\partial}{\partial t}(\varphi C_i) + \frac{\partial}{\partial x}(C_i u) = \Sigma_j a_{ij} R_j \; (i = 1, 2, \ldots. G) \tag{11}$$

where G is the number of the gas species. Gas diffusion was neglected.

$$\text{Conservation of condensed species: } \frac{\partial}{\partial t}(1-\varphi)\rho_k = \Sigma_j a_{kj} R_j \; (k = 1, 2, 3 \ldots. K) \tag{12}$$

where K is the number of solid/condensed species.

$$\text{Energy conservation of gas: } \frac{\partial}{\partial t}(\varphi \rho_g e) + \frac{\partial}{\partial x}(h \rho_g u) = -h_T (T - T_s) + Q_g \tag{13}$$

Energy conservation of

$$\frac{\partial}{\partial t}[(1-\varphi)\rho_s e_s] - \frac{\partial}{\partial x}\left[(1-\varphi) D \frac{\partial T_s}{\partial x}\right] = h_T (T - T_s) + Q_s \tag{14}$$

Using Darcy's law, the gas bulk velocity is given by:

$$u = -\frac{\alpha}{\mu_g}\frac{\partial p}{\partial x} \tag{15}$$

Table 4. Some essential features of reported packed bed models.

Researchers	Dimension and Time Dependence	Heat Transfer			Mass Diffusion	Fluid Flow	Thermo Mechanical Failure	Water Influx	Heat Loss
		Conduction	Convection	Radiation					
Winslow [47]	1D & T	√	√			D			
Thorsness *et al.* [48]									
Thorsness and Rozsa [42]	1D & T	√	√			D			
Khadse *et al.* [46]	1D & PS	√	√			D			
Uppal *et al.* [43]	1D & PS	√	√			D			
Thorsness and Kang [51]	1D & T	√	√		√	D	√		
Gunn and Whitman [45]	1D & PS	√			√			√	
Abdel-Hadi and Hsu [8]	1D & T	√	√		√				

1D = One-dimensional, D = Darcy flow, PS = Pseudo-steady-state, T = Transient.

Table 5. Reaction rate control mechanisms of the main reactions in the gasification process.

Researchers	Drying	Pyrolysis	Char Reaction				Water-Gas Shift Reaction	Gas Phase Reaction		
			$C + O2 \rightarrow CO_2$	$C + CO_2 \rightarrow 2CO$	$C + H_2O \rightarrow CO + H_2$	$C + 2H_2 \rightarrow CH_4$	$CO + H_2O \leftrightarrow CO_2 + H_2$	$CO + 0.5O_2 \rightarrow CO_2$	$H_2 + 0.5O_2 \rightarrow H_2O$	$CH_4 + 2O_2 \rightarrow CO_2 + 2H_2O$
Winslow [47]	D	P	K	K	K	K	K			
Thorsness *et al.* [48]	D	P	K	K	K	K	K	I	I	I
Thorsness and Rozsa [42]	D	P	K	K	K	K	K	I	I	I
Khadse *et al.* [46]		P	K	K	K	K	K			
Uppal *et al.* [43]		P	K	K	K	K	K	I	I	I
Thorsness and Kang [51]			K	K	K	K	K	I	I	I
Gunn and Whitman [45]		EC	P		P					
Abdel-Hadi and Hsu [8]	D	P	P	P	P	P	E			

D = Diffusion-limited, I = Infinite rate, K = Kinetic (power law) and bulk diffusion, P = Power law kinetics, EC = Experimental correlations.

The primary objective of their preliminary works was to predict and correlate reaction and thermal-front propagation rates and product-gas composition as a function of coal bed properties and process operating conditions, as well as the secondary objective of testing the correctness of the physical and chemical model they developed. Despite the simplifying assumptions as shown in Table 6, their results were generally in accordance with steady-state laboratory measurements (Table 7) Experimental run 5 was carried out using back heating in order to eliminate radial heat losses. As a result, model calculations with the assumption of no heat loss were closer to the data of experimental run 5. Although the lab-scale experiment did not represent the field-scale system as the porosity (50%) and the permeability (150 D) were too high and the particle diameter (1 cm) was too small, the agreement between the model calculations and the lab experiments indicates that the simple physical and chemical models of Thorsnesset *al.* [48] are useful to some extent in the design of the gasification process.

Beside the model validation with experimental data, they also illustrated various features of their simulations with time or spatial variations and compared them with the experimental results where possible. The solid and gas temperature distribution along the length in their system at different times indicated the propagation of the reaction and drying/pyrolysis fronts at different speeds (see Figure 3).

Table 6. Input parameters used for model calculation and experimental runs from a 1.6 m reactor.

Input Parameters	Calculated (Thorsness and Rozsa [42], Thorsness *et al.* [48])	Calculated (Winslow [47])	Reactor experiments (Thorsness and Rozsa [42], Thorsness *et al.* [48])	
			Run 4 (no backheating)	Run 5 (with backheating)
Coal Dimension	L = 150 cm, D = 15 cm	L = 150 cm, D = 15 cm	L = 150 cm, D = 15 cm	L = 150 cm, D = 15 cm
Initial Porosity	50%	44.4%	50%	50%
Initial Permeability	150 Darcy		150 Darcy	150 Darcy
Coal Particle Size	1 cm	1 cm	1 cm	1 cm
Injected Gas Flow Rate	2×10^{-4} gmole/cm^2· s	2.1×10^{-4} gmole/cm^2· s	2×10^{-4} gmole/cm^2· s	2×10^{-4} gmole/cm^2· s
Initial Feed Temperature	450 K	450 K	450 K	450 K
Pressure	4.8 Atm	4.8 Atm	4.8 Atm	4.8 Atm
	Injected Gas Composition			
O_2	14.1%	13.4%	13.4%	13.4%
N_2	24.0%	22.0%	22.0%	22.0%
H_2O	83.5%	84.4%	84.4%	84.4%
Steam/O_2 Ratio	6	6	6	6

Table 7. Steady-state temperature and major constituents of outlet gas composition.

Parameters	Calculated (Thorsness and Rozsa [42], Thorsness *et al.* [48])	Calculated (Winslow [47])	Measured (average over run) (Thorsness and Rozsa [42], Thorsness *et al.* [48])	
			Run 4 (no back heating)	Run 5 (with back heating)
Peak Temperature (K)		1250		1200
Speed of burn front	18 cm/h	18 cm/h		18 cm/h
	Product Gas Composition (dry basis)			
H_2	43.2%	43.8%	44.9%	44.6%
CH_4	6.6%	4.0%	6.4%	6.9%
CO	22.0%	22.0%	16.5%	19.0%
CO_2	28.2%	30.2%	33.2%	29.5%

The different fronts that were growing at different speeds associated with the movement of the gasification process were also recognized from the plot of the reaction rate as a function of bed length (Figure 4). Those were the coal drying front, then the coal pyrolysis front, followed by the char reaction front considering moving upstream from the outlet. The double-peak of the water-gas shift reaction was due to the changes in equilibrium limits because of changes in temperature (Figure 4). The gas concentration along the bed length showed that the inlet gas composition remained unreacted until the reaction front (Figure 5). However, after an initial rapid change followed by a more gradual one, a sharp change in gas concentration was observed due to the drying-pyrolysis front. Winslow [47] also observed the same trend.

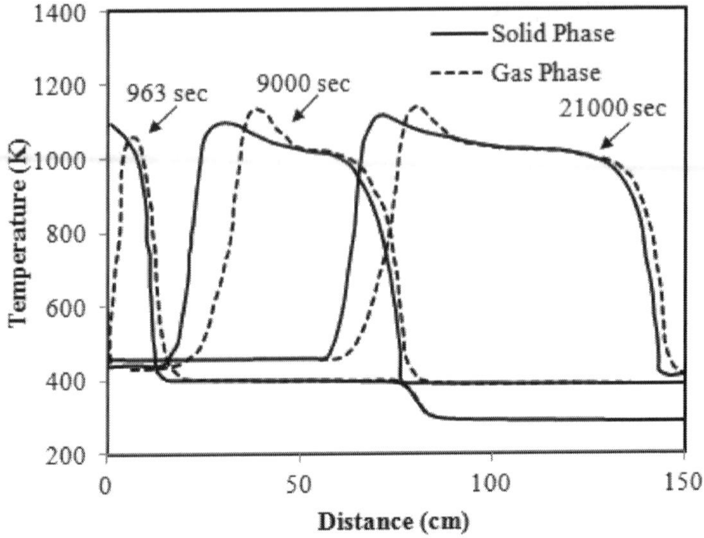

Figure 3. Calculated gas and solid temperature at three different times for the 1.5 m reactor simulation [48].

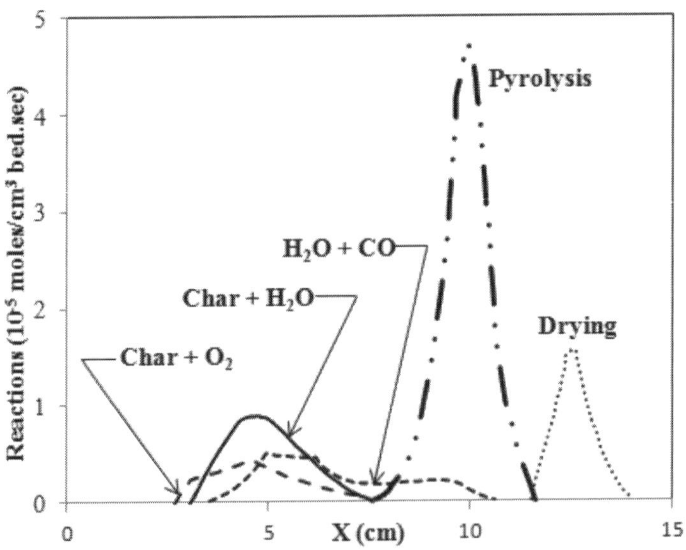

Figure 4. Distribution of principal reaction rates at t = 1 h 13 min [47].

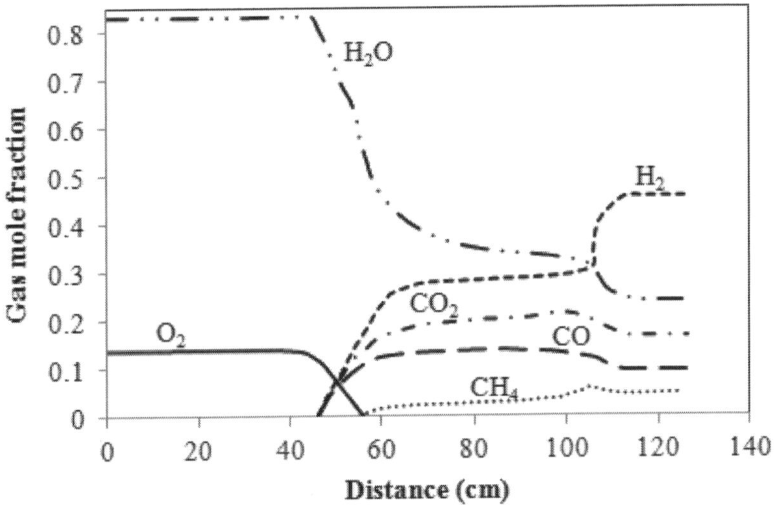

Figure 5. Calculated gas phase concentration at time = 1.4 × 10^4 s for the 1.5 m reactor simulation [48].

A comparison of the position of the calculated reaction and drying front with experimental data showed that the results were in good agreement with the reaction front; however, the predicted drying front movement rate was found higher than the experimental findings (see Figure 6). According to them, the latter observation was due to some uncertainty of the product gas mix and the heat losses occurring during the experiment at the drying zone.

Thorsness *et al.* [48] extended their works by reporting very limited parametric study. For example, with increasing oxygen concentration in the feed gas (holding a constant gas flow rate), they observed an increased gas/solid temperature and the drying rate of the reaction front moved linearly with oxygen concentration. An increased ratio of CO to CO_2 resulting from the increased temperature generation in the system was also detected. With the increase of total dry gas production, a consistent CO/CO_2 ratio was also noticed with the higher oxygen feed. However, for energy recovery per mole of O_2, they indicated an optimum value of oxygen consumption at a mole fraction of approximately 0.2 (see Figure 7). This observation is actually set a limit for the steam-to-oxygen ratio at a value of 4 for maximum energy recovery.

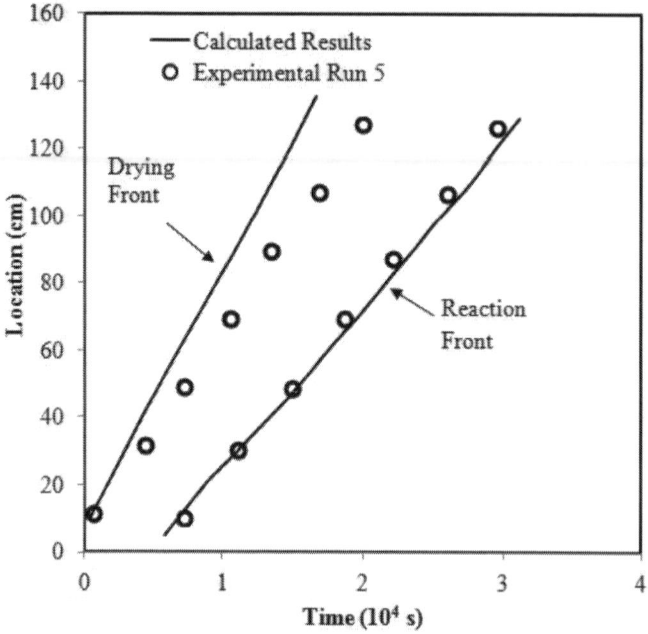

Figure 6. Comparison of calculated and experimental reaction and drying front movements [48].

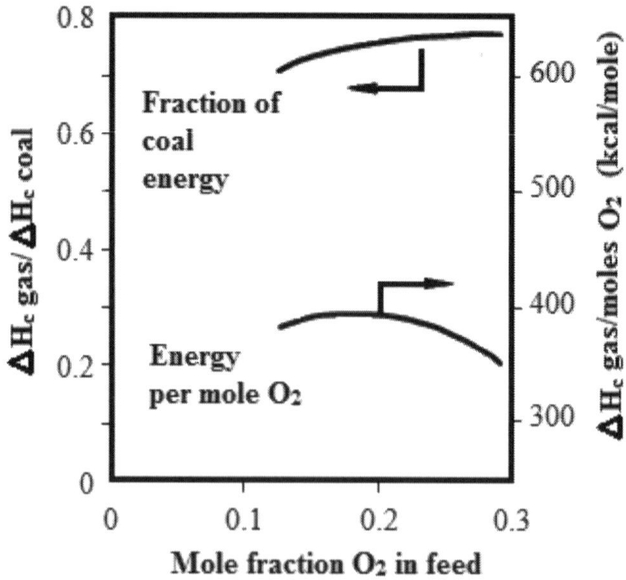

Figure 7. Calculated energy recovery for three levels of oxygen feed concentration [48].

There was an attempt by Thorsness *et al.* [48] to simulate a field-scale system with length equal to 15 m, porosity equal to 0.05, permeability equal to 20 D, outlet pressure equal to 1 atm, and particle diameter equal to 20 cm. However, they could not validate the changes of the process variable except the developing temperature profile that was similar to their lab-scale simulation.

Although they showed time and distance changes in the process variable in the principal direction of gas flow, they did not conduct a detailed sensitivity analysis to investigate the effect of coal particle diameter, porosity, permeability, injection gas flow rate, and other parameters in their model despite the fact that the basis of the model formulation was a 1 cm particle diameter. This imposes the use of their model for specific conditions. The applicability of their model to field conditions where the particle diameter is approximately equal to 20 cm is limited because of the lack of sensitivity of the results to particle size and the validation of field trials. The particle size is particularly important because, in their physical and chemical model, the char reaction rates and other transport coefficients such as heat transfer and mass transfer coefficients are directly or indirectly related to the particle sizes. In addition, because of high initial porosity ($\varphi > 0.25$), they considered a constant permeability in their simulation. There apparently was no use of their formulated functional relationship relating permeability to porosity for considering high initial porosity. The shape of the combustion zone and, hence, the actual gasification volume of the coal could not be determined since the model was only one-dimensional.

Because of its transient nature, the model developed by Winslow [47] can be applied for the cases where rapidly changing transient behaviour is important. However, no stark difference between the model outcomes of transient and quasi-steady-state models was observed, and that is probably why some researchers adopted the quasi-steady-state consideration for ease of calculations.

After three decades, in 2006, Khadse *et al.* [46] developed a simple 1D model following the physical and chemical model as well as the pseudo-steady-state fluid flow model described by Thorsness *et al.* [48]. However, their model differs from the model developed by Thorsness *et al.* [48] in that the drying and mobile water evaporation/condensation reactions were not considered and only coal and char were considered as solid phase. Nevertheless, their work gets attention as they analyzed the effects of various operating and model parameters on the temperature and gas phase and solid compositions in UCG that were not completely incorporated in the models discussed above. Their input parameters for the base case were slightly different than that of Thorsness *et al.* [48] as can be seen from Table 8. However, at least for one run, they simulated the experimental conditions of Thorsness *et al.* [43] in order to compare their calculated exhaust gas composition with the experimental data

(Figure 8). However, they could not compare the calculated results of Thorsness *et al.* [48] that were very close to the experimental results quantitatively, although they followed the development of the model of Thorsness *et al.* [48]. This is possibly due to considering certain parameters that lack specific information in the literature model and neglecting the drying and mobile water evaporation/condensation reactions that Thorsness *et al.* [48] considered. However, their result can be considered only in a qualitative agreement with the experiments of Thorsness *et al.* [48].

Table 8. Comparison of input parameters of Khadse *et al.* [46] with experimental run 5 of Thorsness *et al.* [48].

Parameters	Khadse *et al.* [46] (Base Case)	Experimental Run 5 [48]
	Input Parameters	
Coal Dimension	L = 100 cm, D = 15 cm	L = 150 cm, D = 15 cm
Initial Porosity	20%	50%
Coal Particle Size	1 cm	1 cm
Injected Gas Flow Rate	2×10^{-4} gmole/cm^2· s	2.1×10^{-4} gmole/cm^2· s
Initial Feed Temperature	430 K	450 K
Pressure	4.8 Atm	4.8 Atm
Injection Gas Composition		
O_2	15.4%	13.4%
N_2	7.6%	22.0%
H_2O	77.0%	84.4%
Steam/O_2 Ratio	5	6

The simplicity of the model enabled authors to investigate the effects of various parameters such as O_2concentration, steam/O_2 ratio, and inlet pressure in their model. They found that increasing oxygen concentration (holding the steam fraction constant) in the feed increases the propagation rate of the reaction front and the temperature in the reaction zone. Similarly, increasing the steam fraction (holding the oxygen concentration constant) increased the propagation rate of the reaction front; however, a decrease in temperature was observed. In the absence of nitrogen, they found an optimum outlet product with a steam/oxygen ratio of 1.5. They also found that input parameters did not influence the pyrolysis reaction. They also reported the variation of the gross calorific value of the exhaust gas, on a dry basis, at different pressures with time (Figure 9) where the gross calorific value was calculated using the following formula obtained from Harker and Backhurst [52]:

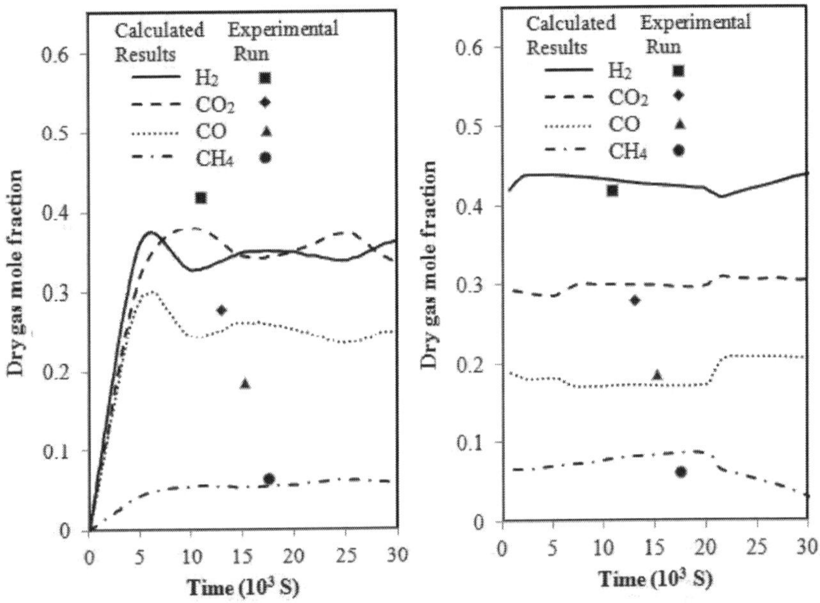

Figure 8. Comparison of calculated exhaust gas compositions with experimental results from Thorsness *et al.* [48]; (**a**) Khadse *et al.* [46]; (**b**) Thorsness *et al.* [48]).

$$Calorific\ Value\left(\frac{kJ}{mole}\right) = H_i \times y_i \tag{16}$$

where H_i is the heat of the combustion of the gas "i" (kJ/mol), and y_i is the mole fractions of the gas "i" in the product gas on a dry basis. However, the calculation was based on the exit concentration of CO, H_2, and CH_4 only. Pressure was found to have no effect on the process (Figure 9). In all the cases, a steady flow rate and composition in the outlet was reached after 5000 s of process simulation.

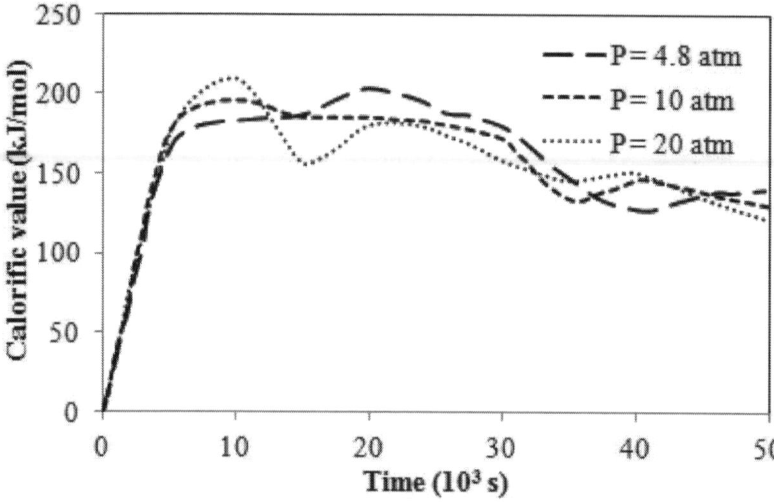

Figure 9. Gas calorific value at different total pressures with time [46].

Recently, Uppal *et al.* [43] also developed another 1D packed bed model adopting the existing model of Thorsness*et al.* [48] with some modifications in the model structure and solution strategy. In order to observe their model capability, they considered a bed length of 4.8 m only and observed various features of their simulations with qualitative agreement with the literature. However, for validating their model with the experimental data, they considered the experimental data from their pilot plant which had dimensions of 150 m × 125 m that contained an array (7 × 6) of wells connected by a network of pipes. For comparing with the model calculation, only two consecutive wells that were 25 m apart from each other, were considered. Table 9 shows their input parameters, some of which are significantly different from the model that they followed.

Table 9. Input parameters for UCG packed bed model developed by Uppal *et al.* [43].

Parameters	Value
Length of Reactor	25 m
Coal Type	Lignite B
Injected Gas	Air
Injected Gas Flow Rate	2×10^{-4} gmole/cm^2·s (for observation of model capabilities)
Initial Feed Temperature	430 K
Pressure	6 Atm
Initial Coal Density	1.25 g/cm^3

In their pilot plant, the injected gas (air only) flow rate was varied with time (Figure 10) and the resultant heating value and the exhaust gas composition were recorded. In their model, they calculated the exit gas heating value and composition utilizing the experimental inlet gas flow rate. In addition, they used a nonlinear optimization technique in order to compensate the uncertainty in coal and char by optimizing the composition parameters of some pyrolyzed products (H_2, CH_4, and char). The calculated results were in a reasonable agreement with the experimental results (Figure 11a,b). However, according to them, a better prediction can be obtained by increasing the optimization variables. Their model lacks some detailed information such as particle diameter, porosity, permeability, *etc.*, which are considered important for understanding a UCG process. Nevertheless, their model indicates that a control strategy can be employed by manipulating the inlet flow rate to maintain a desired heating value in the presence of disturbances.

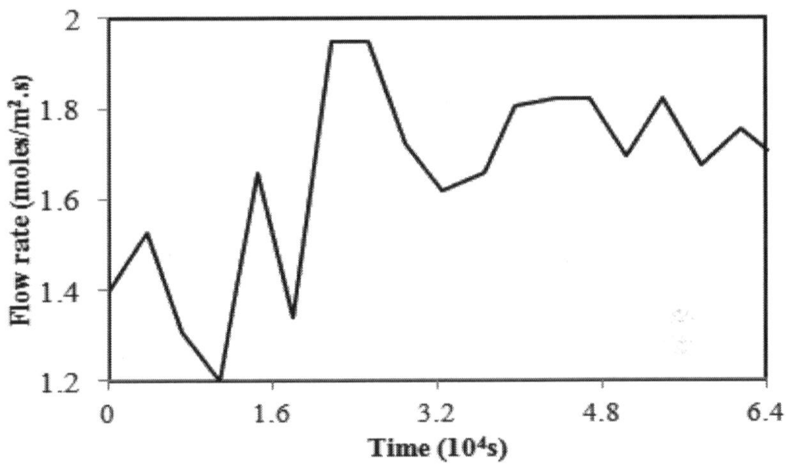

Figure 10. Flow rate of the injected air with time [43].

In 1986, Thorsness and Kang [51] developed a generalized 2D model for describing reacting flows through the packed bed using the following governing equations:

Mass balance:

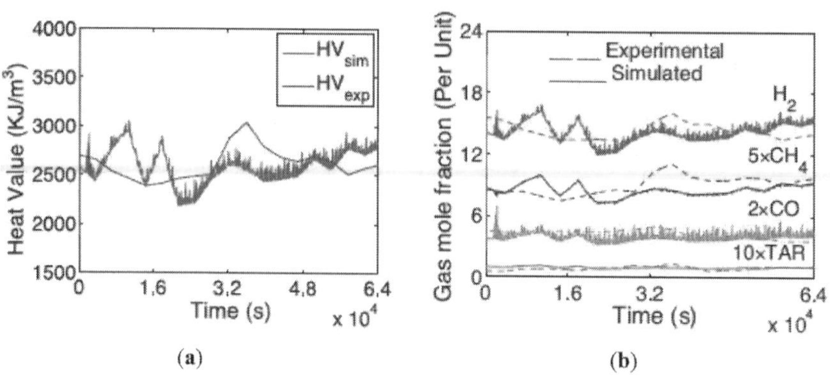

Figure 11. Experimental and simulated (**a**) heating values and (**b**) mole fractions of important gases of the product gas mixture with time (reproduced from Uppal *et al.* [43]).

Overall gas conversion:

$$\frac{\partial(\varphi C)}{\partial t} = -\nabla.(\overline{v}\varphi C) + \Sigma_j a_{ij} R_j + S_m \; (i = 1, 2, \ldots. G) \tag{17}$$

Conservation of gas species:

$$\frac{\partial}{\partial t}(\varphi C_i) = \nabla.(\overline{v}.\varphi C_i) + \nabla.[C D_i \nabla y_i] + \Sigma_j a_{ij} R_j + S_{m_i} \tag{18}$$

Overall solid conservation:

$$\frac{\partial}{\partial t}[\rho_s(1-\varphi)] = -\nabla.[(1-\varphi)\rho_s \overline{v_s}] + \Sigma S_{s_k} \tag{19}$$

Solid species conservation:

$$\frac{\partial}{\partial t}[\rho_s Y_{s_k}(1-\varphi)] = -\nabla.[\rho_s \overline{v}_s Y_{s_k}(1-\varphi_e)] + S_s \tag{20}$$

Energy balance:

They assumed an identical gas and solid temperature and incorporated one energy equation for the combination of gas and solid as follows:

$$\frac{\partial}{\partial t}\left[\varphi\Sigma(C_i h_i) + (1-\varphi)\rho_s\Sigma(Y_{s_k} h_k)\right]$$
$$= -\nabla.[\Sigma h_i \overline{J_i}] - \nabla.[\rho_s \overline{v_s}\Sigma(Y_{s_k} h_k)] + \nabla.[k_{eff}\nabla T] + \Sigma rate_i \nabla H_i \tag{21}$$

Considering the negligible difference between the gas and solid temperature obtained by the earlier models [47,48], the assumption of identical gas and solid temperature seems to be justified.

In their model, they incorporated diffusion effects, wall transport, and also char combustion and water-gas shift reaction rates based on Shell Progressive (SP) and Ash Segregation (AS) reaction models. In the SP model, a core of unreacted solid was assumed to be surrounded by a shell of ash through which the gas phase reactants diffuse (Figure 12a). On the other hand, the AS model assumed continuous exposure of unreacted material to the gas stream due to the ash segregation (Figure 12b).

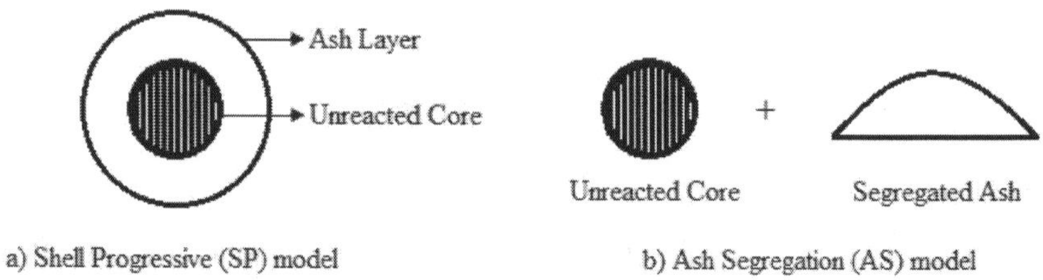

Figure 12. Different reaction regimes in packed bed model.

Reaction rates for these two models were as follows:

$$\frac{1}{R_{SP}} = \frac{1}{R_{bulk\ mass\ transfer}} + \frac{1}{R_{ash-layer\ diffusion}} + \frac{1}{R_{diffusion\ into\ +\ ntrincis\ reaction}}$$
$$\frac{1}{R_{AS}} = \frac{1}{R_{bulk\ mass\ transfer}} + \frac{1}{R_{diffusion\ into\ +\ ntrincis\ reaction}} \tag{22}$$

Because of the generalized nature of their model, they tested various cases (*i.e.*, transient concentration and thermal wave, steady heat transfer phenomena, *etc.*) for non-reacting packing materials through the packed bed and compared them favourably with other available studies. Although their generalized model is 2D, only one case (steady heat transfer phenomena) was solved using the 2D model. For all other cases, the 1D model was considered. For validation of a UCG model, they calculated gas composition, temperature, and carbon fraction considering a steady 1D model with very limited gas species and compared the results with the analytical data obtained from literature. They analyzed the following three disparate situations related to the UCG process which were not considered in the earlier models:

-Case 1: Transient water injection into the midpoint of the packed bed during gasification.

-Case 2: Coal wall drying using a hot gas flow.

-Case 3: Wall regression during gasification.

Table 10 shows the input parameters for the above situations. For the first case, seven gas species (N_2, O_2, H_2, CO, CO_2, water vapor, and CH_4) and eight reactions (reactions 3–11) were considered. However, drying and pyrolysis reactions were ignored. During gasification process, when the product gases reached a reasonably steady state, water was injected at the midpoint of the bed, slightly downstream of the reaction front, and continued until the mole fraction of CO and CO_2 were observed to approach steady values. However, the liquid water influx was assumed to turn into steam instantly with a flow rate of 8×10^{-4} mol/s. A sudden change in CO and a gradual change in H_2 were noted after the water injection; however, the gases undergoing changes returned to pre-injected conditions as seen inFigure 13.

Table 10. Input parameters for the three situations considered by Thorsness and Kang [51].

Input Parameters	Case 1	Case 2	Case 3
Packed Bed Dimension	L = 150 cm, D = 5 cm	L = 25 cm, D = 10 cm	L = 1 m, D = 1 m
Initial Gas Temperature	900 K	900 K	900 K
Initial Bed Temperature		300	
Gas Composition			
H_2O	74%	100% (or N_2)	66.6%
O_2	26%		33.3%
Injected Gas Flow Rate	1.5×10^{-3} mole/s	8×10^{-3} mole/s (1 mol/s-m^2 of bed)	6 moles/s

For case 2, a uniformly permeable non-reacting bed with a water-saturated wall at steam temperature was assumed through which hot gas was passed. It was observed that the time to reach steady state increased rapidly with a decreasing flow rate as if the average wall-drying rate is a function of the injected gas flow (Figure 14). However, the wall-drying rate was found to be limited by the available energy for too low injected hot gas due to the insufficiency of the total heat injected.

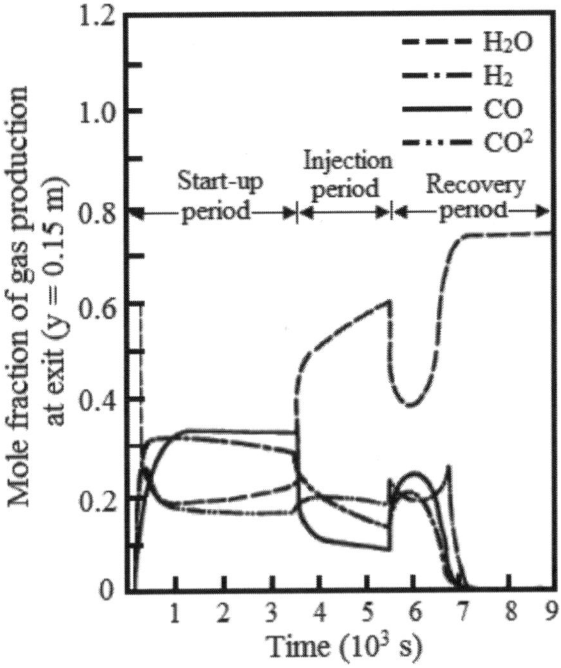

Figure 13. Product gas changes during gasification with midpoint water injection [51].

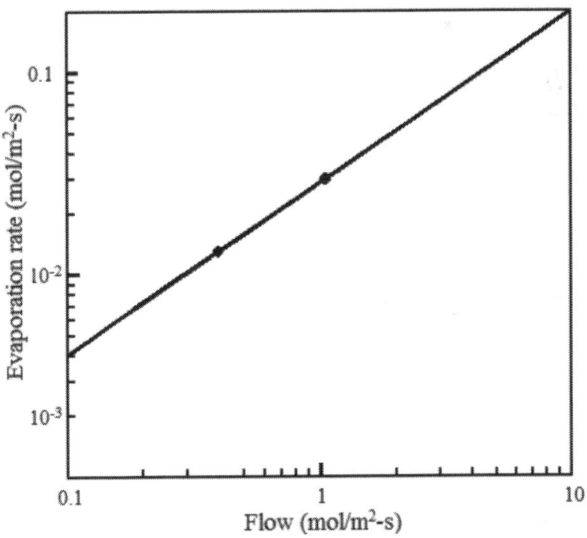

Figure 14. Drying rate *vs.* flow rate (log-log plot, from Thorsness and Kang [51]).

For case 3, the model system was considered to be filled with rubble material consisting of ash in the center and char near the walls and at the top. The walls were considered as coal that can proceed with gasification reactions. The objective of this model system was to determine the thickness of the char bed at the wall that would lead to a self-sustaining system. Calculations were performed using all seven gas species. The computed wall regression rate (*i.e.*, the carbon production rate) was found nearly constant with the increasing char layer thickness, while the carbon consumption varied linearly with the char thickness (Figure 15). A wall char bed thickness of 7.5 cm was found close to a self-sustaining system, since the rate of carbon consumption was nearly equal to the carbon production. However, the calculated average wall regression rate of 7.7×10^{-7} m/s (0.07 m/day), when comparing the characteristic field result (~0.5 m/day), indicates that their model is missing some additional physics.

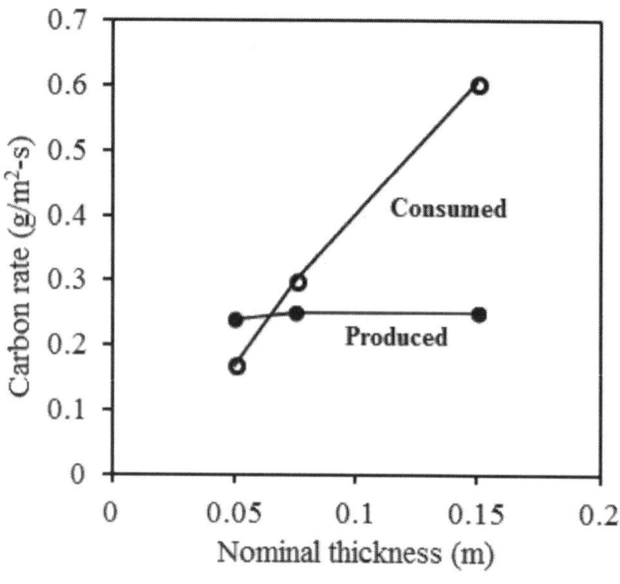

Figure 15. Carbon production rate *vs.* wall char layer thickness [51].

In 1976, in parallel to LLNL, Gunn and Whitman [45] also developed a 1D linear mathematical model separately for UCG. Their work is particularly important as their model is the only packed bed model cited in the literature that was compared with the field trials (Hanna, Wyoming) and exhibited a fairly good agreement with the experimental data (Table 11). However, field test data was considered at a time span when a steady-state condition was observed in order to conform closely to the assumption of their model. In addition to the gas composition, temperature, and heating value, they also determined the ratio of the gas production rate to the gas injection rate, the effect of reservoir water influx on the heating value of gas, the distribution of thermal energy and thermal efficiency during the gasification process that closely matched the experimental observations and, thus, provided enhanced understanding of the UCG process.

Table 11. Comparison of experimental and calculated data for an air injection rate = 1631 Mcf/day [45].

Components	Composition (Mole Percent)	
	Field Test Results	Calculated Data (Average Value)
H_2	18.57	18.60
CH_4	4.10	4.92
N_2 + Ar	47.81	47.13
CO	16.35	15.83
CO_2	12.33	13.29
H_2S	0.04	0.07
Ethan +	0.80	0.14
Gas Production Rate (Mcf/day)	2728	2734
Gas Production Temperature, °F	493 (measured at the surface)	533 (measured at the bottom of the well)
Maximum Temperature, °F	1871	1951
Heating Value, Btu/scf	170.6	164.6
Thermal Efficiency (%)	89.2	87.4

Unlike the above models, the water-gas shift reaction and methanation were not included in their model. Methane was considered to be produced solely through pyrolysis. Only two reactions, water-gas and combustion reactions, were considered. The reaction rate of the water-gas reaction was adopted from the work of Gadsby *et al.* [53]; however, the constants were determined from the gasification of the pitch coke. For the combustion reaction, the reaction data of Lewis *et al.* [54] was used that was obtained from the combustion of metallurgical coke. Even the correlation of the thermal conductivity they used was not intended for subbituminous coal. Drying was not considered in their model. Despite this and the large errors in the reaction rate, they observed only negligible errors in the model predictions.

In their model, water vapor, carbon monoxide, hydrogen, carbon dioxide, oxygen, inert gases (nitrogen and argon), and devolatilized material from the coal were considered. For convenience, the volatile products resulting from the pyrolysis of coal were treated as a single pseudocomponent; however, the pseudocomponent was broken down into individual gases for the purpose of calculating total gas composition. From the proximate and ultimate analysis of the experimental subbituminous coal (from the Hanna field test), they also formulated a correlation between the weight fraction of volatile matters and the temperature with high accuracy. In order to obtain the composition of devolatilized gas, they considered the experimental devolatilized data of the Hanna I coal seam at 900 °C along with the assumption that the tars

(16.2%) and light oil (0.9%) and water (13.3%) in that analyses cracked to the stoichiometric proportion of methane, carbon monoxide, and hydrogen. Their assumption led to the composition for devolatilized products listed in Table 12.

Table 12. Composition of devolatilized products assumed by Gunn and Whitman [45].

Components	Percent
H_2	42.26
CO	28.53
CO_2	3.79
CH_4	24.36
H_2S	0.37
Ethane +	0.69
Total	100
Average molecular wt.	14.76

Their model transformed the set of partial differential equations into a set of ordinary differential equations by considering the pseudo-steady-state approximation. They considered nine integrated continuity equations (two for overall solid and mass balances and others for individual gas species), a partially integrated energy equation, and a macroscopic material balance equation from which a constant combustion front velocity could be calculated iteratively. The assumption of the constant combustion velocity seems to be valid, according to Figure 6, with what was observed by Thorsness *et al.* [48].

Their model predicted the exhaust dry gas composition and obtained a close agreement with the experimental gas composition history with the exception of methane (Figure 16). However, the deviation of methane was believed to be a result of considering a high concentration of methane in the devolatilized products. The only parameter fitted in their model from the field trial was the water influx in order to get a good prediction for the data that interferes with the water influx rate. Their calculated results were also closely matched with experimental data for ratio of gas production/air injection and gross heating value (Figure 17 and Figure 18). However, the declination of those quantities with time was believed to be the result of the gradual increase of water flux during field test. They calculated the heating value of gas by changing the ratio of water influx to air injection rate and observed a maximum heating value at a ratio of about 0.15. Usually the maximum heating value at a particular ratio of water intrusion to air injection rate implies the control strategy of closely maintaining the optimum value either by controlling the water intrusion rate by pressure of any other means or by changing the air injection rate.

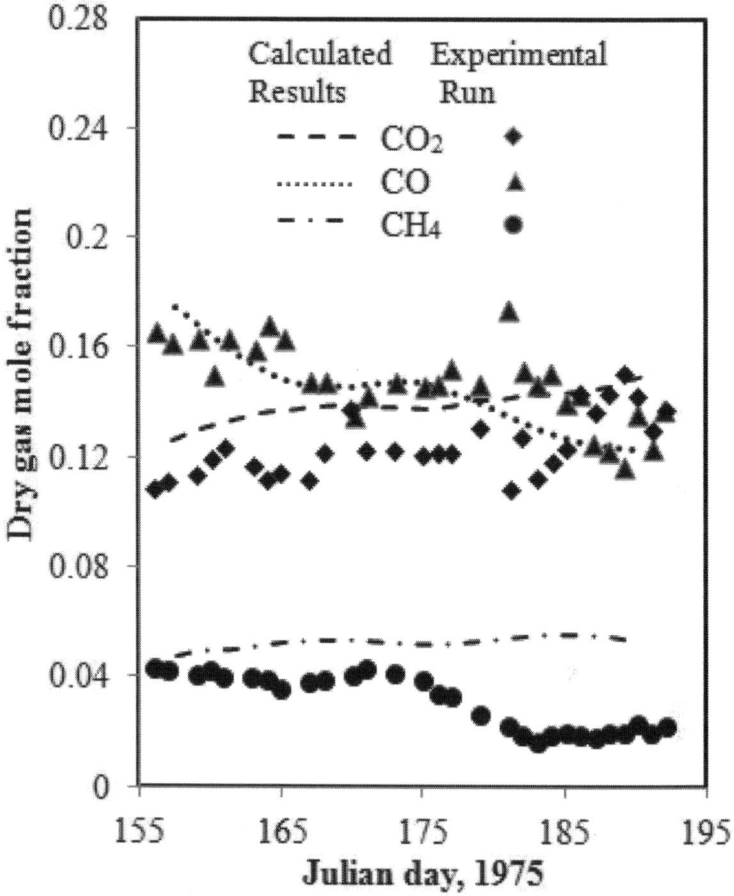

Figure 16. Comparison of calculated exhaust gas compositions from Gunn and Whitman [45] with experimental results from the Hanna field trial.

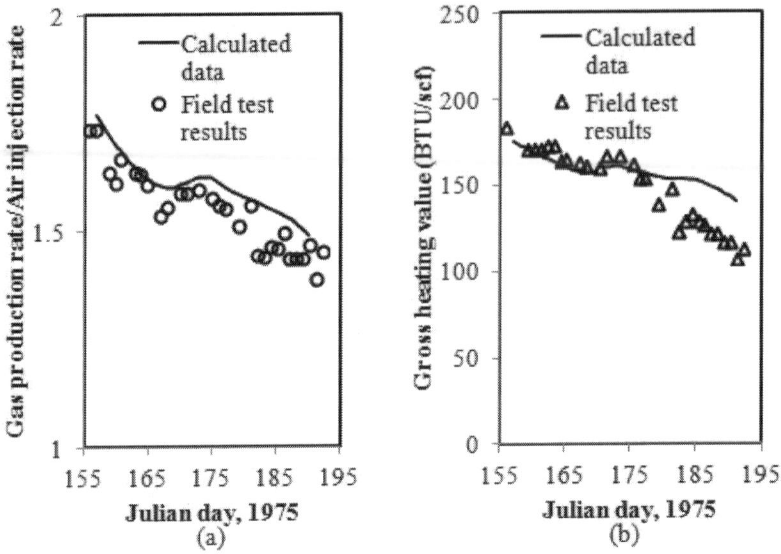

Figure 17. Comparison of calculated data and field test: (**a**) gas production/air injection ratio and (**b**) gas heating value [45].

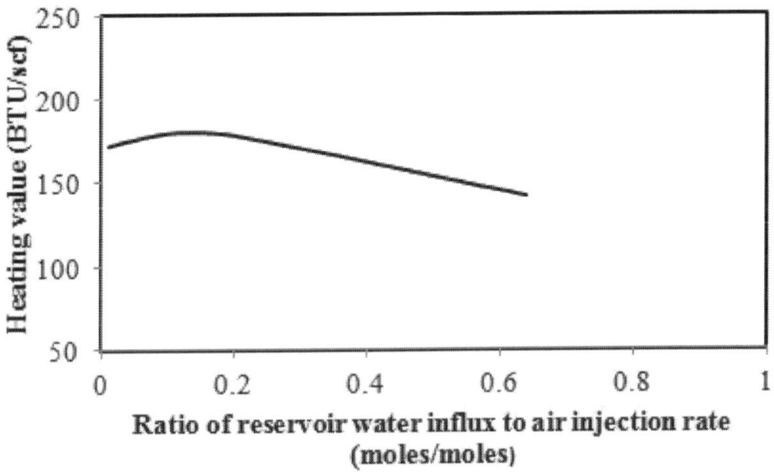

Figure 18. Effect of reservoir water influx on heating values of gas [45].

The calculated values of the distribution of thermal energy released by coal combustion were also in good agreement with the experimental values (Table 13). From the experimental data, it was observed that only a small amount of energy from a 30-foot coal seam was lost to the overburden and base rock. This implies that neglecting heat losses to the surrounding rocks does not hamper a model outcome too much. The last column of Table 13 shows a higher heating value of the product gas when injected air was preheated at 480 °F. In order to get higher economic value and energy conservation, they suggested using a part of the heat of vaporization of the produced water to heat the inlet gas. They also reported that in all cases, the reaction zone was very narrow, generally two feet or less. This fact was verified by both model calculations and by temperature measurements in the observation wells.

Table 13. Distribution of thermal energy from *in situ* coal combustion [45].

	Calculated, %	Field Test, %	Calculated with Preheating, %
Produced Gas			
Gross heating value	87.4	89.1	91.3
Sensible heat....	5.8	5.2	2.9
Heat of vaporization of water vapor in produced gas	3.4	3.3	2.4
Heat Loss to Surrounding Rock	1.3	3.8	1.3
Heat Loss in Ash	2.1	0.6	2.1
Total	100	100	100

Although they obtained a good agreement with the experimental results for a number of phenomena, they could not quantitatively predict the variation of the temperature and mole fraction along the bed length with the experimental value because of the poor consideration of the thermal conductivity, heat capacities, and the uncertain accuracy of the reaction rates. However, they obtained a qualitative agreement with the experimental values. Despite very close agreement with the experimental data, their model is not sufficient for predicting all aspects of gasification as their model neglected some important phenomena, such as drying, CO_2 gasification, the water-gas shift reaction, gas phase reactions, *etc.* However, because of the low water/air ratios that were experienced with the Hanna tests, there might be little or no water-gas shift reactions. As a result, the Gunn model was not hampered by neglecting this reaction. In their model, there is also a lack of sensitivity analysis for the kinetic parameters they used. In addition, they included water influx data from a particular field trial. However, water influx is a highly variable phenomenon that depends on coal seam properties and hydrology. All these lacking parameters limit their model for general applicability in the UCG process.

Abdel-Hadi and Hsu [8] extended previous models by developing pseudo-2D geometry with a moving burn front in the axial direction. A rectangular domain with a length of 1.5 m and width of 1 m was used in their model. Their governing equations are nearly similar to the equation considered by Winslow [47] and Thorsness and Kang [51]; however, they included carbon consumption in the reaction zone in order to track the burn front, and the equation is given by:

$$\rho_s w_c v_c = 12 \int_0^{x_{rz}} \left(\Sigma_j R_{c-ij}\right) dx \qquad (23)$$

where w_c is the carbon mass fraction, v_c is the velocity of the combustion front, and x_{rz} is the total length of the reaction zone. They also performed an immobilization transformation of coordinates in order to formulate the irregular front motion as follows:

$$\xi = \frac{x - X(y,t)}{H_1 - X(y,t)} \qquad \eta = \frac{y}{H_2} \qquad (24)$$

All other governing equations were transformed to these new coordinates (Figure 19). This allowed them to calculate the progressive configuration of the gasified zone at various stages of gasification (Figure 20) which, in turn, facilitated the calculation of the rate of coal seam consumption (Figure 21). The conversion rate of the coal seam is found fairly constant. In order to gain confidence for this model, they have compared their model with the laboratory results reported by Thorsness and Rozsa [42] and obtained a good agreement with the experimental data.

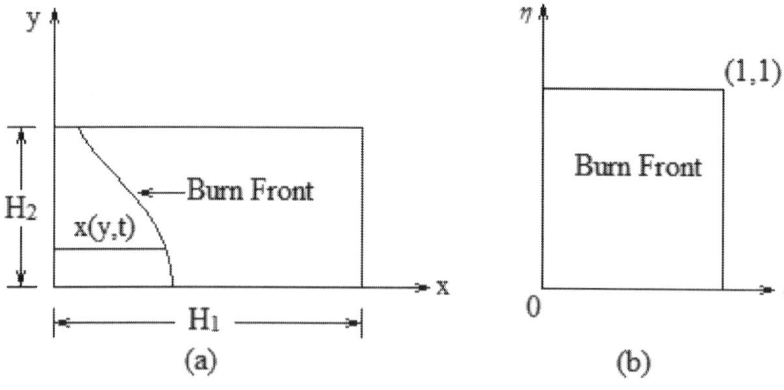

Figure 19. Schematic description of the problem: (**a**) conventional coordinate system; (**b**) immobilization coordinate system (redrawn from Abdel-Hadi and Hsu [8]).

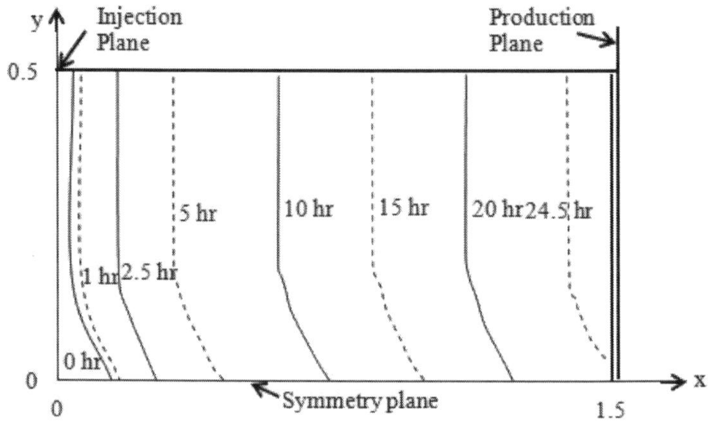

Figure 20. Progressive configurations of combustion fronts during gasification [8].

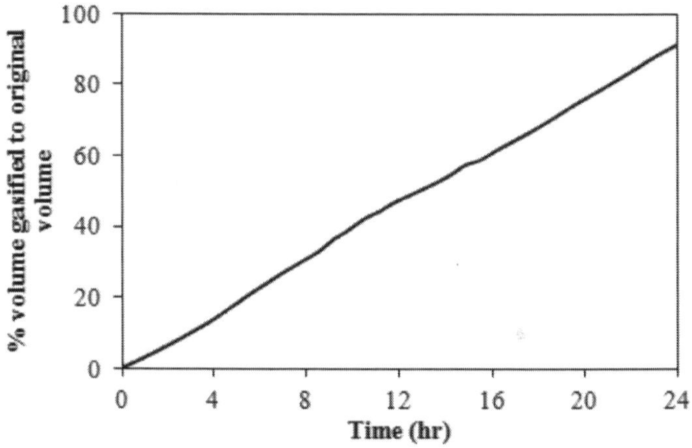

Figure 21. Consumption of coal during gasification [8].

The packed bed models have been validated with laboratory experiments to some extent. These models exhibit good agreement for the gas composition. In order to calculate the heat recovery and gas composition, the model can be very effective. However, the extension of these models to field trials is infeasible, since other cavity growth mechanisms such as thermo-mechanical failure could not be incorporated into the models. Also, as pointed out by Winslow [47], this method requires a fine grid at the vicinity of the reaction front that limits its applicability to three-dimensional field-scale trials. However, recent advancement of computational power can easily overcome this limitation.

2.2. Channel Models

Beside packed bed models, a number of channel models have also been developed in the first decades of modeling. The channel model assumes that coal is gasified only at the perimeter of the expanding permeable channel [19]. In this approach, the UCG process is represented by an expanding channel when two distinct zones of rubble/char and open channel exist. This approach is considered due to the observation of the formation of the open channel structure after the gasification phase is terminated in different field tests of coal seams [55,56]. Figure 22 shows the basic concept and physics behind this approach: "Air or oxygen flows down the central channel and is convected by turbulent flow to the boundary layer along the channel wall. The oxygen diffuses through the boundary layer to the solid surface and reacts. The hot combustion gases diffuse back through the boundary layer to the channel" [40]. The channel model is more useful for analyzing sweep efficiency [19].

Table 14 and Table 15 represent a glimpse of a list of channel models discussed here with their essential features and reaction rate control mechanisms of the main reactions in the gasification process, respectively.

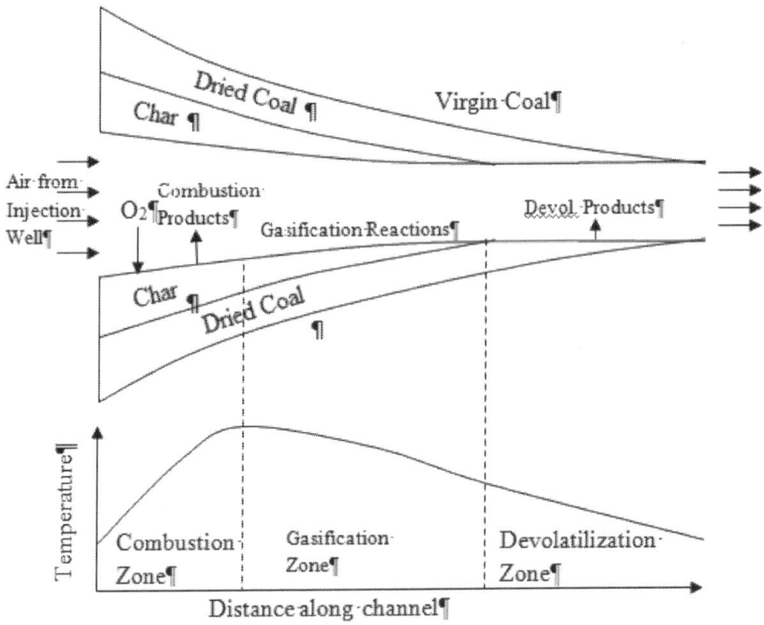

Figure 22. Reactions and transport phenomena in channel model (redrawn from Gunn and Krantz [40]).

Table 14. Some essential features of reported channel models.

Researchers	Dimension and Time Dependence	Heat Transfer: Conduction	Heat Transfer: Convection	Heat Transfer: Radiation	Mass Diffusion	Fluid Flow	Thermo Mechanical Failure	Water Influx	Heat Loss
Dinsmoor *et al.* [44]	2D and T	√	√	√		P		√	
Eddy and Schwarz [57]	1D and T	√	√	√	√	M	√	√	√
Luo *et al.* [58]	2D and T	√	√		√				
Batenburg *et al.* [55]	1D and SS	√	√	√		D			
Pirlot *et al.* [59]	2D and S	√	√		√	D		√	
Kuyper [56,60]	2D and T	√	√	√	√	NS			
Perkins and Sahajwalla [61]	2D and T	√	√	√	√	NS			

D = Darcy flow, P = Plug flow, M = Mixed flow, NS = Navier-Stokes, S = Steady State, SS = Semi-Steady State, PS = Pseudo-Steady State, T = Transient.

Table 15. Reaction rate control mechanisms of the main reactions in the gasification process.

Researchers	Drying	Pyrolysis	Char Reaction: $C + O2 \rightarrow CO_2$	Char Reaction: $C + CO_2 \rightarrow 2CO$	Char Reaction: $C + H_2O \rightarrow CO + H_2$	Char Reaction: $C + 2H_2 \rightarrow CH_4$	Water-Gas Shift Reaction: $CO + H_2O \leftrightarrow CO_2 + H_2$	Gas Phase Reaction: $CO + 0.5O_2 \rightarrow CO_2$	Gas Phase Reaction: $H_2 + 0.5O_2 \rightarrow H_2O$	Gas Phase Reaction: $CH_4 + 2O_2 \rightarrow CO_2 + 2H_2O$
Dinsmoor *et al.* [44]			K	K	K			P	P	
Luo *et al.* [58]			P	P	P	P	P	P	P	P
Batenburg *et al.* [55]				E	E	E	E		E	
Pirlot *et al.* [59]				E	E	E	E	E	E	E
Kuyper [56,60]				D				M		
Perkins and Sahajwalla [61]				P				M		

D = Diffusion limited, E = Equilibrium, K = Kinetic (power law) and bulk diffusion, M = Turbulent mixing limited, P = Power law kinetics

Dinsmoor *et al.* [44] developed a steady-state, 1D channel model by assuming that the gasifier behaves as an expanding cylindrical cavity in the coal seam with reactions taking place at the walls. In their model, for simplicity, no pyrolysis reactions were considered. Heat transfer included conduction for solids only; however, both convection and radiation were included between the wall and gas. Axial heat conduction in the gas phase was neglected. They also considered water influx as evenly distributed along the length of the tube. Char reactions (reactions 3–5) and two gas phase reactions (reactions 9 and 10) were considered in their model. Because of slow channel evolution, they incorporated a pseudo-steady-state assumption for changes associated with the gas phase. They treated a fully developed flow process with the channel despite the initial cylindrical channel diameter (0.3 m) being much smaller than the channel length (60 m). Considering forced convection as the dominating mechanism of mass transfer, they simulated coal gasification with a forced convection mass transfer correlation. For heterogeneous reaction kinetics, they considered the surface reaction rate constant and wall diffusion resistances. For wall diffusion resistances, the mass transfer coefficients were calculated from a standard correlation for the turbulent flow of gases in tubes. Inlet gas was assumed to be air at 330 K and 6.9 atm with no water added. Their predicted gas compositions and temperature profile showed clear evidence of the presence of a separate oxidation and reduction zone; however, a very high temperature (2400 K) was noted near the inlet (Figure 23). According to them, this high temperature was due to neglecting heat loss at the end of the tube and including the radiation heat transfer. However, it appears that apart from their explanations, neglecting the pyrolysis reaction could be another reason for getting high temperature initially, as during ignition, a significant amount of heat is used to create char by the pyrolysis of coal.

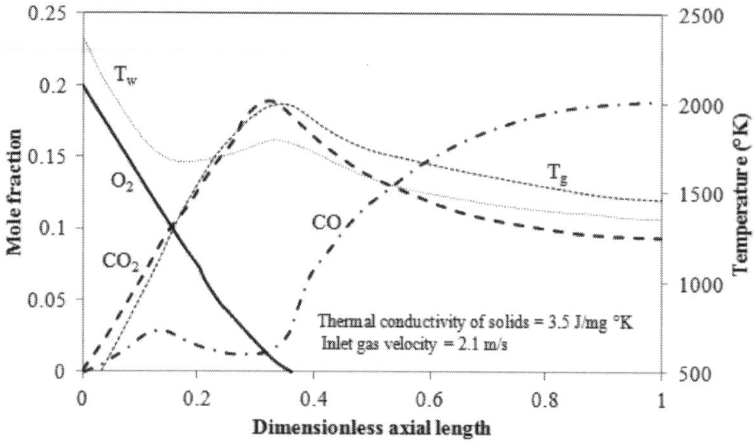

Figure 23. Typical channel temperature (gas and wall) and gas composition profiles without water influx [44].

The quality of the predicted product gas observed was inferior to the gas quality usually obtained in the packed bed model for similar situations. For a constant blast velocity, the reaction rates were observed to decrease with the evolution of the channel due to the constant mole fraction of the oxygen. As a result, the oxidation zone became longer which, in turn, was responsible for increased heat losses and the deterioration of gas quality. Compared to the packed bed combustion front (approximately 0.2 m), the oxidation zone (approximately 20 m) was much longer. However, these observations were not supported by any field observations. Apparently, because of the long oxidation and reduction zones, they concluded that "a successful UCG system cannot be operated in the channel regime".

Almost at the same time, the above conclusion was negated by the work of Schwartz *et al.* [62] as they found an increase of mass transfer by several orders of magnitude when natural convection in channels was considered as the controlling mechanism of mass transfer instead of forced convection alone. The intense mixing of the injected blast with the boundary layer gases depends on the Grashof number (Gr) of the natural convection flow. A good mixing begins when the Grashof number exceeds 10^8 and becomes intense at about 10^{10} [63]. However, for dominant natural convection flow, the gases within the boundary layer produce a circulatory flow within the cylinder and a good mixing is obtained. However, Schwartz *et al.* [62] assumed two different modes of the circulatory pattern based on coal seam thickness as depicted in Figure 24. Figure 25 shows the difference of cavity growth with and without considering the natural convection. Later, it was established by mathematical correlation that the models assuming forced convection mass transfer as the controlling mechanism for cavity growth were not born out of field trials [15].

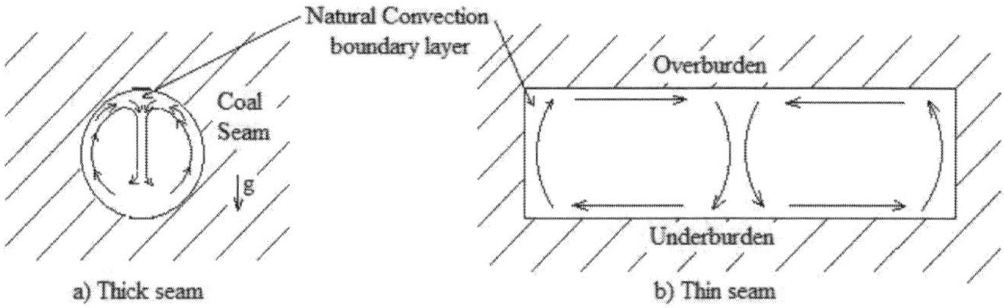

Figure 24. Assumed circulation patterns in the thick seam and thin seam configurations (redrawn from Schwartz *et al.* [62]).

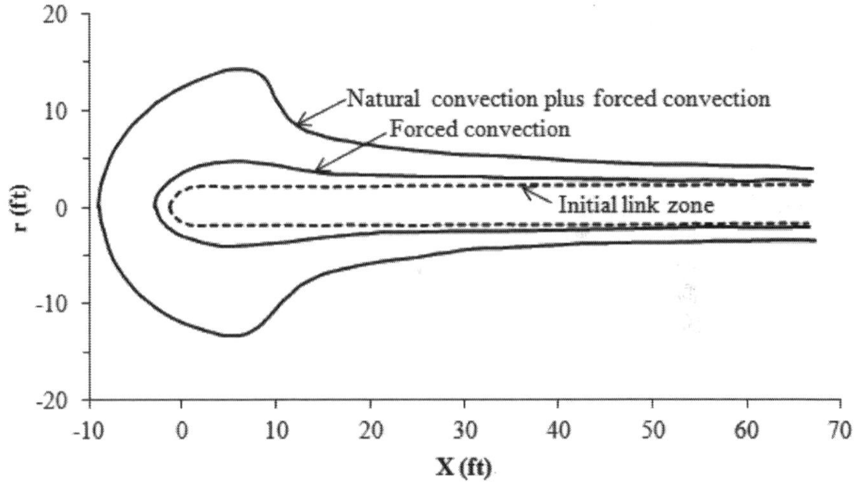

Figure 25. A comparison of cavity growth via forced convection and natural convection [62].

Schwartz *et al.* [62] were the first investigators to consider the natural convection as the controlling mechanism of heat and mass transfer in UCG cavities. In their later paper, Eddy and Schwarz [57] developed a 2D model and described the evolution of the cavity based on the movement of the cavity wall. The blast from the injection well was transported through the reacting walls of a constant temperature via the convection process and ultimately diffused to the wall and reacted with the coal. The oxygen diffusion rate was calculated by natural and forced convective heat and the mass coefficient. The roof and floor of the cavity were assumed to be non-reacting surfaces. They considered neither subsidence nor rubblization. They included the water-gas shift reaction and two other gas phase oxidations along with three char reactions (reactions 3–5) as considered by Schwartz *et al.* [62]. In order to capture the experimentally proven teardrop shape of the cavity formation, they divided the cavity growth into different regions, *i.e.*, a hemispherical region in the vicinity of the injection well and a series of non-equal diameter cylinders (step cylinder) downstream from the injection well, according to the flow process and a cylindrical link zone (Figure 26). At any location, the cavity was assumed to be a cylindrical cross-section until the cavity diameter was equal to the seam thickness, which is when the model switches to a rectangular cross-section. It was assumed that all coal volume located in the hemispherical region was consumed in the reaction with oxygen. The volume of char was calculated from the mass of char burned due to the combustion and the density of the char. The half-body volume was then fit to the calculated char volume. Flow was treated as an entrance region of pipe, as the calculated entrance length was much larger than the vertical well distance. The entrance length was calculated using the following equation:

$$\frac{L_e}{d} = 0.0288 Re_d \tag{25}$$

where L_e is the entrance length and d is the pipe diameter. The minimum Reynolds number in the system was 4000, which resulted in an entrance length of 48.8 m. The entrance length was considered to increase with the increase of the cavity diameter. According to them, as fresh blast proceeds down the axis and products of combustion are released at the walls, the forced and natural convection results in a swirling flow that ultimately helps mix the flows and enhance the flow of oxygen to the vicinity of the wall where boundary layer convective diffusion dominates. The transport of heat and species inside the cavity was considered to be governed by empirical correlations for the turbulent transport phenomena in enclosures. The model was able to reproduce the results of the Hanna II and Pricetown field trials qualitatively.

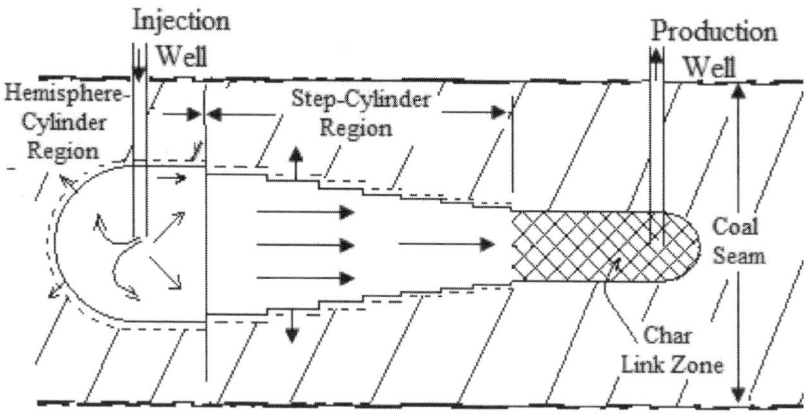

Figure 26. Conceptual sketch of the linked vertical well computer model geometry (redrawn from Eddy and Schwarz [57].

Recently, Luo *et al.* [58] extended the Schwartz *et al.* [57] model by including heat transfer and more coal wall and gas phase reactions (reactions 3–10). Flow inside the cavity was solved based on irrotational fluid flow inside an enclosure, which describes velocity potential based on geometric features of the enclosure. The geometry is shown inFigure 27. The stream function and velocity potential were calculated using the following equations:

Steam function:

$$\Psi = \frac{\dot{m}}{4\pi}\cos(\pi-\theta) + \frac{Ur^2}{2}\sin^2(\pi-\theta) \tag{26}$$

Velocity potential:

$$\varnothing_v = \frac{\dot{m}}{4\pi r} + Ur\cos(\pi - \theta) \tag{27}$$

where U is the velocity of the uniform stream and $\dot{m}$ is the mass flow rate. Pressure distribution in the cavity was determined using Bernouli's equation:

$$p + \frac{1}{2}\rho u^2 = p_\infty + \frac{1}{2}\rho U^2 \tag{28}$$

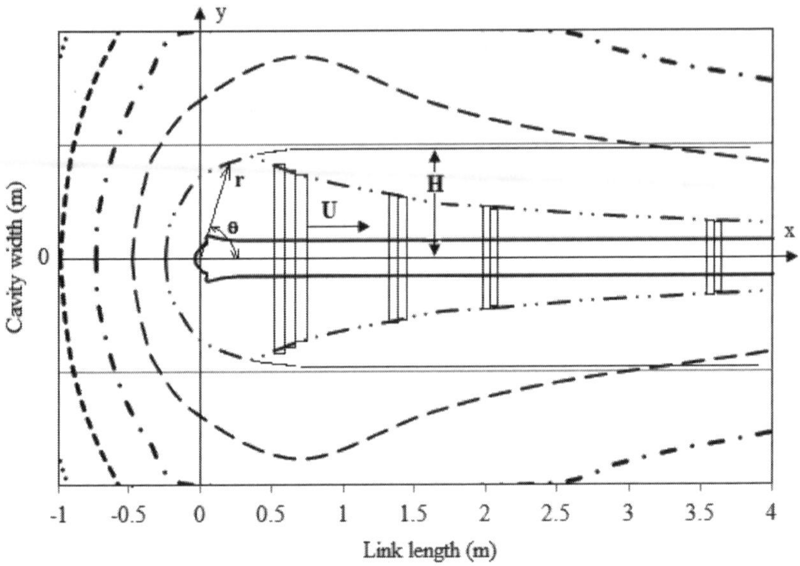

Figure 27. UCG cavity defined in the work of Luo *et al.* [58].

In the Luo model, Fluent was used to predict the cavity shape at different times based on heterogeneous reaction rates at the cavity wall. In order to quantify the turbulent intensity, they used the standard k-ε model for the turbulent kinetic energy and dissipation. For heterogeneous and homogeneous reactions, they used kinetics/the diffusion control process and a finite rate/eddy dissipation model, respectively. Their model was validated with Chinchilla field trials. The calculated coal consumption closely matched the Chinchilla trial field data (Figure 28). For sweep cavity geometry, their model was further validated with the data from the Hanna II and III UCG trials. There is less than 5% error between the results generated with the model (Hanna II: 2436.6 tons of coal gasified in 25 days; Hanna III: 4139.3 tons in 38 days) and reported data (Hanna II: 2500 tons of coal gasified in 25 days; Hanna III: 4200 tons in 38 days). Their model predicts a hemispherical shape for the cavity geometry; however, the model is limited since heat and mass transfer characteristics of the cavity are unknown. Also, the coupling of this model with the mechanical failure of the coal would be cumbersome, since the accumulated rubble in the cavity changes the transport phenomena inside the cavity.

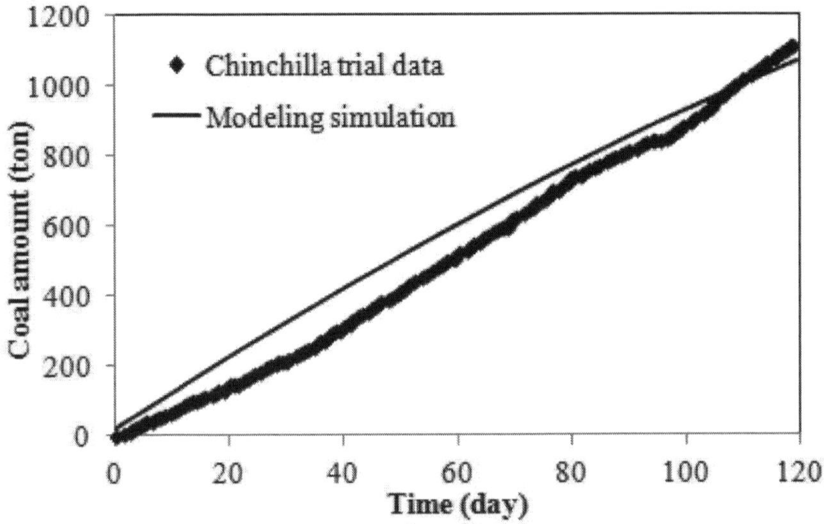

Figure 28. Coal consumptions for modeling and Chinchilla trial data [58].

Batenburg *et al.* [55] developed a semi-steady-state 2D model for UCG in open channels for the developed gasifier only. Unlike other models, they assumed that oxygen instantaneously reacts with the combustible gases present in the channel instead of reacting with the coal surface. Their interest was only to investigate the process within the channel after the injection gas percolates through the inert permeable rubble zone. Reaction rates were calculated based on resistances in the system including boundary layer, pore diffusion, surface phenomena, and chemical kinetics. For laminar and turbulent natural convection along a smooth vertical wall, they used the Sherwood number (Sh) that is correlated with the Grashof number (Gr) and Schmid number (Sc) as follows:

$$Sh = 0.59\ Gr^{1/4}Sc^{1/4}, \qquad Gr < 10^9 \qquad \text{(laminar)} \tag{29}$$

$$Sh = 0.13\ Gr^{1/3}Sc^{1/3}, \qquad Gr > 10^9 \qquad \text{(turbulent)} \tag{30}$$

For forced convection-dominated flow the Sherwood number was expressed in terms of the Reynolds number (Re) as follows:

$$Sh = 0.027\ Re^{4/5}Sc^{1/3} \tag{31}$$

Heat transfer was modeled by radiation between the walls of the channel. They also included the effect of natural convection due to the temperature difference. Their results showed that the effect of pressure on gas composition is negligible (Figure 29). Their results also indicated that natural and forced convection transfer coefficients are in the same order of magnitude and both of them are important.

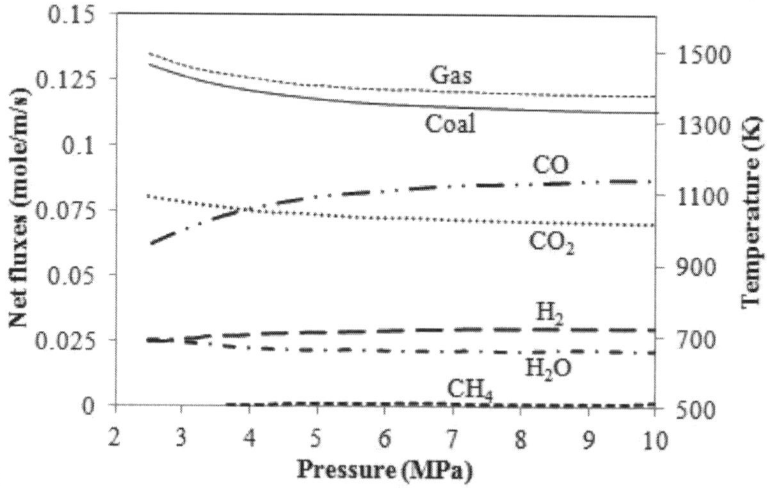

3

Figure 29. Net fluxes and temperature as a function of process pressure for oxygen injection rate = 0.1 mol/m/s and water injection rate = 0.05 mol/m/s [55].

Pirlot *et al.* [59] developed a 2D steady-state model by extending the idea of Batenburg *et al.* [55] with two distinct zones: a low-permeability rubble and ash around the injection point and a high-permeability peripheral zone along the coal wall (Figure 30). After a short transitory starting phase, those two zones were identified by other researchers [64,65] for a thin seam at great depth. During initial combustion, a cavity was identified partially filled with inert materials near the injection hole [64,65]. Their simulation for cavity evolution was based on one main parameter, the permeability ratio between the low- and high-permeability zones. The gasifying agent was assumed to pass through the low-permeability zone surrounding the injection point prior to its arrival in the high-permeability zone, where reactions with the coal wall occur. They combined two separate models: (1) a flow model for the calculation of the flow through the low-permeability porous zone using Darcy's law and the continuity equation; and (2) a chemical model for the calculation of the chemical processes occurring between gas and the coal wall in the high-permeability zone using empirical correlations for mass and heat transfer for the packed bed by assuming plug flow in the gas phase. The coal consumption rate was calculated only on the channel wall. Their model did not consider the details of moisture and volatile matter released by drying and pyrolysis. They assumed that volatile matter is released in the form of water and hydrogen in proportion to the consumption rate of carbon. They concluded that the permeability ratio is one of the main parameters for determining the success of underground coal gasification because of the observation of the increasing final gasifier area, power, and trial duration with the increase of the permeability ratio. According to them, the cavity growth and shape obtained from their model were in reasonable agreement with the Pricetown field trials.

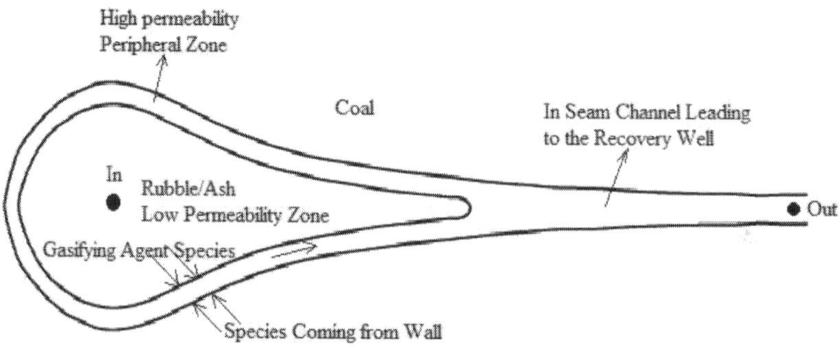

Figure 30. Conceptual sketch of the two separate zones and species flow direction (redrawn from Pirlot *et al.* [59]).

Kuyper [56,60] developed a 2D model to describe UCG process in a cross-section of an open channel (Figure 31) for a typical western European coal layer of thin seams (1–2 m). Field trials of UCG indicated the growth of the cavity upwards and radially outwards around the injection well as gasification proceeds [66]. As a result, at thin seams, the top wall is exposed to rock materials and failure of the overburden is apparently expected. Such a failure would create open channels around the rubble materials deposited on the floor just about the injection hole; as shown in Figure 31. Considering this fact, in their model, the top and bottom walls were assumed as impermeable and adiabatic rock materials. However, the main focus of their work was to obtain an insight for heat and mass transfer due to double-diffusive turbulent natural convection in which both the temperature and concentration gradients play a role in the transport process. The justification of using double-diffusive turbulent natural convection was based on the expectation that $Gr/Re^2 >> 1$, considering the Grashoff number (Gr) of the natural convection flow in the process is approximately 10^{10} while the Reynolds number (Re) of the forced convection flow is about 10^4. In this situation, if the forced convection flow is superimposed on the natural convection flow, the heat and mass transport phenomena inside the cavity will be dominated by double-diffusive natural convection and a spiral flow can be expected. The justification for this assumption can be found elsewhere [67]. The Navier-Stokes equation and the k-ε turbulence model were used to describe fluid flow due to large gradients of density and concentration in the channel. Radiation and convection were assumed to be the major heat transfer mechanisms in the channel. Model results showed that oxygen is consumed far from the coal wall by combustible gases. The double-diffusive natural convection in the channel was observed to cause the periodic generation and collapse of CO_2 bubbles (Figure 32). Also, mass transfer was reported to be the controlling mechanism (the rate-limiting step) for reduction reactions at the coal wall. They also studied the effect of CO_2 injection into the coal seam and reported that CO_2 injection has a similar effect as adding steam, although to a lesser extent.

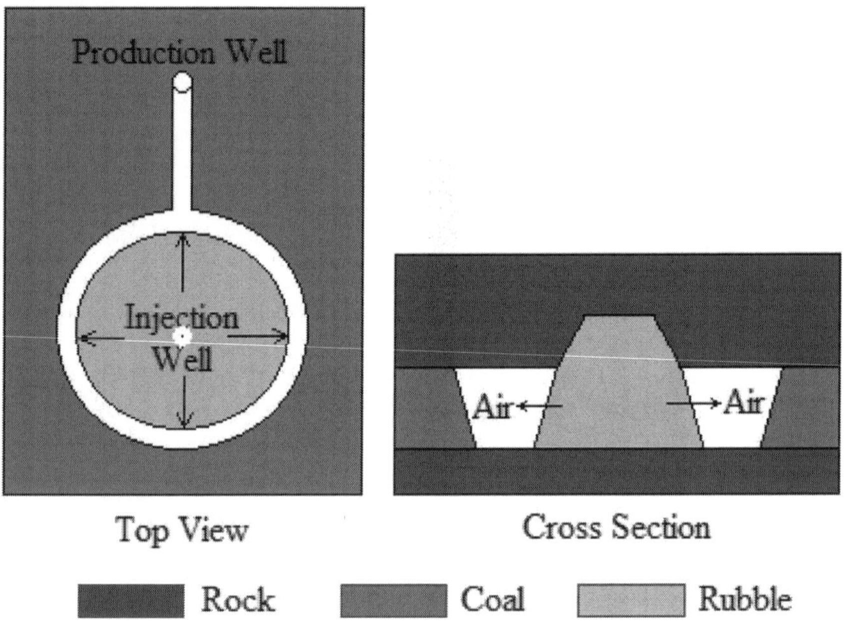

Figure 31. Channel formation in thin seams (Modified from Kuyper *et al.*[60]).

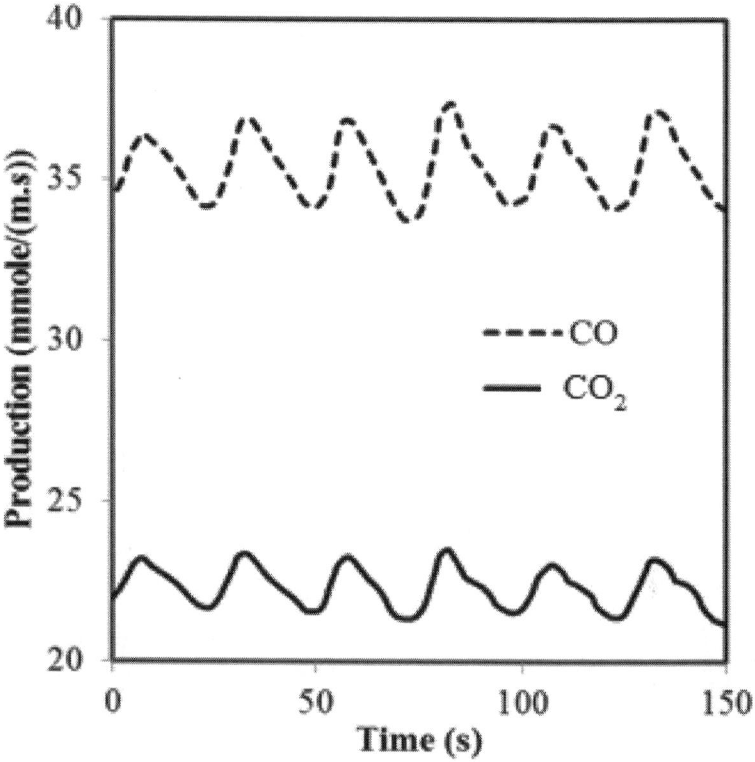

Figure 32. Production of CO and CO_2 as function of time for κ = 0.12 $m^{-1}$and an oxygen injection rate of 47 mmole $m^{-1}s^{-1}$ [60].

Because of the identification of UCG application in thick coal seams and the presence of a porous bed of ash overlying the injection point and a void space above the UCG cavity based on field trial excavation [66,68,69] as depicted in Figure 33, Perkins and Sahajwalla [61] considered a thick coal seam and expanded Kuyper's model by including an ash layer at a lower part of the channel (Figure 34). They developed a 2D axisymmetric model by using Fluent to investigate double-diffusive natural convection along with relevant reactions in a partially filled cavity. Although the bottom wall was considered to be inert and adiabatic, the side and top walls were considered to be carbon due to the thick coal seam. The temperature of the gas and the porous ash bed were assumed to be in thermal equilibrium. The flow though the porous ash bed was modeled using a laminar approximation; however, the turbulence model is only applied to the void region of the cavity. The low Reynolds number turbulence model of Launder and Sharma [70] was employed for the turbulent model by replacing the constants in the standard *k*-ε model with the function of the turbulent Reynolds number in order to extend the model's applicability to low-turbulence regions that are close to walls. The following turbulent kinetic energy (k_t) and the turbulent dissipation rate (ε) equations were used in order to account for turbulent intensity:

$$\frac{\partial}{\partial t}(\rho_g k_t) + \nabla.(\rho_g \overline{v} k_t) = \nabla.\left[\left(\mu_g + \frac{\mu_t}{\sigma_{k_t}}\right)\nabla k_t\right] + G_{k_t} + G_b - \rho_g \varepsilon + Y_m + \dot{w}_{k_t} \tag{32}$$

$$\frac{\partial}{\partial t}(\rho_g \varepsilon) + \nabla.(\rho_g \overline{v} \varepsilon) = \nabla.\left[\left(\mu_g + \frac{\mu_t}{\sigma_\varepsilon}\right)\nabla \varepsilon\right] + C_{1\varepsilon}\frac{\varepsilon}{k_t}(G_k) - C_{2\varepsilon}\rho_g \frac{\varepsilon^2}{k_t} + \dot{w}_\varepsilon \tag{33}$$

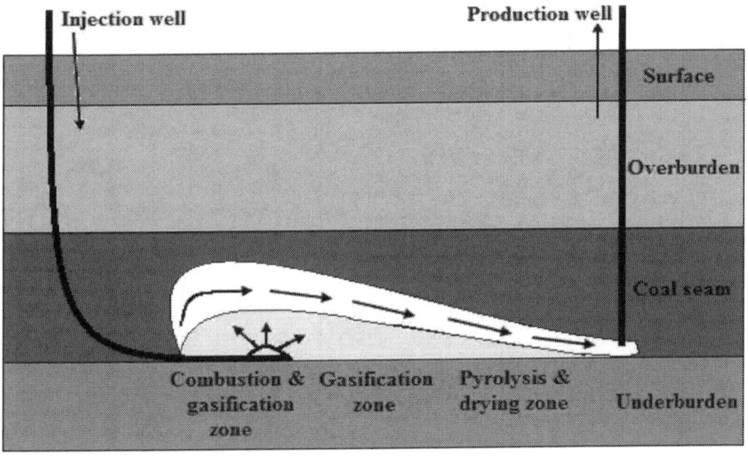

Figure 33. Schematic of an underground coal gasification cavity from field observation (redrawn from Perkins and Sahajwalla [61]).

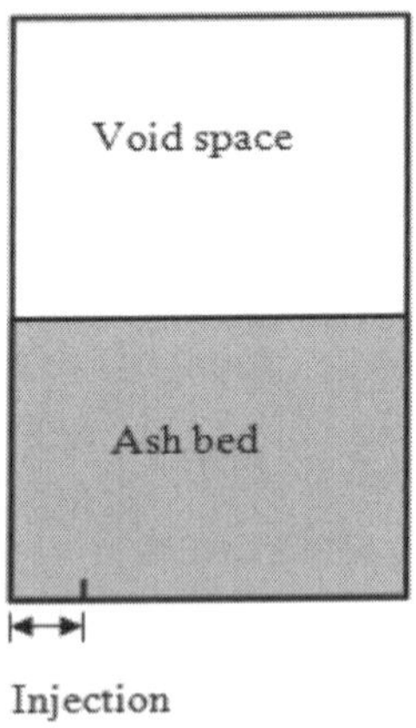

Figure 34. Channel geometry in Perkins and Sahajwalla [61].

For other conservation equations for void space, they followed the standard governing equations developed for porous media with the exception of φ (porosity) = 1. For the simplicity of their model, only three chemical reactions (reactions 2, 8, and partial char combustion) were considered. They found that the flow behaviour in the void space is dominated by a single buoyant force due to the temperature gradient resulting from the combustion of oxygen with CO produced from the gasification of CO_2 at the walls. From the observable entity, by changing the injection point, they concluded that oxygen should be injected at the bottom of the channel, otherwise valuable gasification products would be oxidized, leading to a low heating value of the production gas. The model was verified with Biezen's experiments [71] for double-diffusive natural convection in a trapezoidal channel.

Considering the importance of the flame front propagation in the gasification channel for viable operations in UCG, Saulov *et al.* [72] developed a simplified physical model to describe the primary features of the flame behaviour in a long channel through a coal block. According to their model, a number of factors, such as air flow rate, flame temperature, oxidizer diffusion rate, and radiative heat transport, *etc.*, are the decisive factors to determine if the flame tends to propagate toward the injection point (reverse combustion) or the downstream direction (forward combustion). The flame speed is affected by the air flow rate through the energy balance in the channel. Their model predicts the flame propagation toward the injection point (upstream) for a low air flow rate and toward the downstream direction for a high air flow rate. These predictions from their model are in agreement with qualitative observation by Chernyshv [73]. The flame speed is affected by the air flow rate through the energy balance in the channel. The introduction of oxygen instead of air is found to increase the flame temperature and reactions rates and, as a result, the flame propagates at a faster rate. They also reported a quantitative comparison with the experimental data reported in the book by Skafa [20]. Figure 35 demonstrates two burn velocities, *i.e.*, upstream end velocity, s, and downstream end velocity, s_b, as functions of the injection air speed in the upstream undisturbed section of the channel. It is noted that positive values of the velocity mean that the flame front propagates downstream, while negative values imply that the front moves upstream.

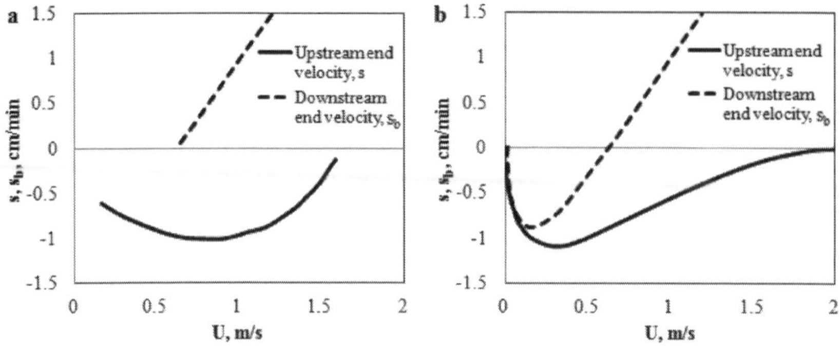

Figure 35. Dependence of burn velocity on the air flow rate [72]: (**a**) experimental data from Skafa [20], and (**b**) calculated data from Saulov *et al.*[72].

In the case of very hot and diffusion-controlled flame, Saulov *et al.* [72] considered the limit of high temperatures, a high activation energy, and a strong air flow. Under these conditions the surface of the channel is considered to have two zones, cold and hot. The temperature is found insufficiently high in the cold zone to initiate reactions, while in the hot zone any oxygen on the surface exhausted quickly due to instant reactions. These zones are separated only by a very small distance due to high activation energy. The overall reaction rate is determined by the rate of diffusion of oxygen to the hot zone, while the oxygen concentration on hot walls is essentially nil. Under such conditions the flow in the gasification channel is turbulent and the turbulent flame is fully controlled by diffusion. The injection rate is found to have no control over the flame position. On the other hand, if the combustion of coal begins with devolitalization reactions at low temperatures and these reactions play a noticeable role in initiating the rest of the oxidation process or in the overall energy balance, the flame position is affected by the air speed and becomes controllable. On the other hand, if the devolitalization reactions during the combustion of coal begin at low temperatures and play a noticeable role in initiating the rest of the oxidation process or in the overall energy balance, the flame position is affected by the air speed and becomes controllable. In this case, the flame is not very hot and the cooling effect of the air flow is strong.

The consideration of natural convection is found to be one of the main phenomena in the channel model development. Natural convection plays an important role for the mixing of injected blast gas and the gases coming from the channel wall. The channel model is found to better calculate sweep efficiency. However, most of the channel models neglected drying and pyrolysis which are considered very important in the coal block model. In order to determine cavity shape and size, the channel model is preferred.

2.3. Coal Slab Models

Some models have attempted to describe the UCG coal seam as a coal slab. These models describe the process by movement of various defined regions in a coal slab perpendicular to the flow of the injected blast gas. These regions usually include the gas film, ash layer, char region, dried coal, and virgin coal. The existence of different regions is due to the slow heating rate of the UCG. At a very high heating rate, there is a possibility of the coincidence of a drying front with a combustion front [74]. The general framework of these models is shown in Figure 36. Tsang [74] was the first to use this approach considering the observation of the development of drying, pyrolysis, and gas-char reactions zones around the most permeable path in the coal seam. In addition, the profiles of temperature and saturation as well as the direction of heat and mass transfer exhibited from the pyrolysis experiment (Figure 37) of a small coal block performed by Westmoreland and Forrester [75] inspired him to consider necessary assumptions, such as the existence of a drying front, heat flux, and mass transfer direction, *etc.*, for this approach. In that experiment, a constant heat was applied to the coal surface and career gas was supplied along the length of the cylinder. Some researchers considered the virgin coal region as a wet zone due to the presence of water which is separated from the dry zone by the evaporation/drying front. The evaporation of water is assumed to take place entirely at the retreating drying front. Thus, the model following this approach must be a moving boundary value problem with phase change which is known as the "Stefan" problem. In this approach, as can be speculated from Figure 36, there is an efflux of steam, devolatilized materials, and "self-gasification" products from the wall to the cavity, while there is a counter-current flux of heat towards the cavity wall. "Self-gasification" is considered as the reaction of the gases (steam and devolatilized gases) with hot char while they pass through it before flowing into the cavity. Because of the consideration of different layers, unlike other types of models, for each layer separate mass and energy balances are usually considered. As a result, the governing equations for mass and energy balances are of split boundary types. The mass flux is considered to be diffusion dominant. Table 16 and Table 17 represent a glimpse of a list of coal slab models discussed here with their essential features and reaction rate control mechanisms of the main reactions in the gasification process, respectively.

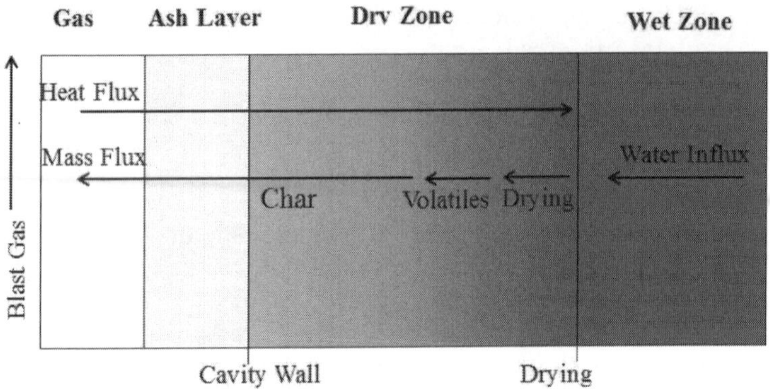

Figure 36. Scheme of coal block models (modified from Perkins and Sahajwalla [76]).

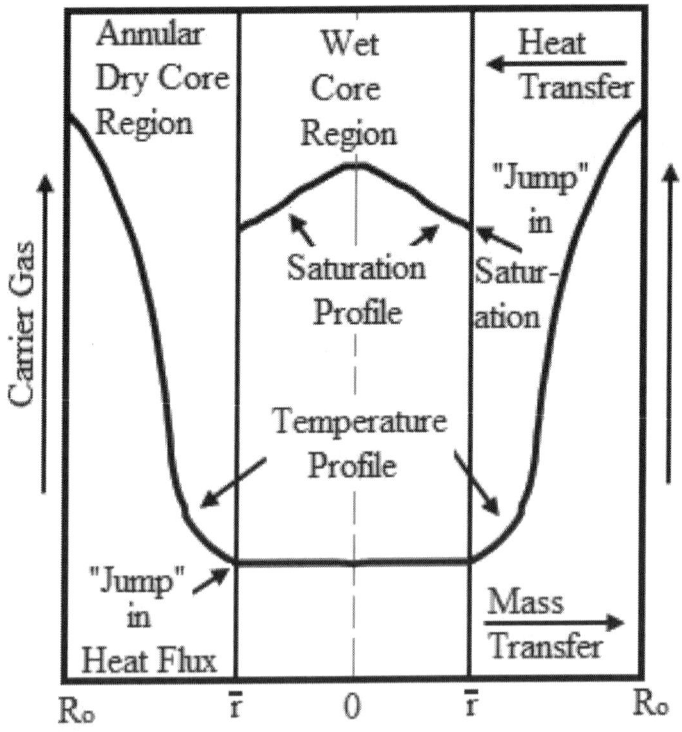

Figure 37. Schematic diagram of temperature and saturation profile and heat and mass transfer in coal block (redrawn from Tsang [74]).

Table 16. Some essential features of reported coal slab models.

Researchers	Dimension and Time Dependence	Heat Transfer			Mass Diffusion	Fluid Flow	Thermo Mechanical Failure	Water Influx	Heat Loss
		Conduction	Convection	Radiation					
Tsang [74]	1D and T	√	√	√	√			√	
Massaquoi and Riggs [77,78]	1D and S	√	√			D			√
Park and Edgar [79]	1D and T	√	√			D			
Perkins and Sahajwalla [76,80]	1D and PS	√	√	√	√	NS	√	√	

D = Darcy flow, NS = Navier-Stokes, S = Steady State, T = Transient, PS = Pseudo-steady.

Table 17. Reaction rate control mechanisms of the main reactions in the gasification process.

Researchers	Drying	Pyrolysis	Char Reaction				Water-Gas Shift Reaction	Gas Phase Reaction		
			$C + O_2 \rightarrow CO_2$	$C + CO_2 \rightarrow 2CO$	$C + H_2O \rightarrow CO + H_2$	$C + 2H_2 \rightarrow CH_4$	$CO + H_2O \leftrightarrow CO_2 + H_2$	$CO + 0.5O_2 \rightarrow CO_2$	$H_2 + 0.5O_2 \rightarrow H_2O$	$CH_4 + 2O_2 \rightarrow CO_2 + 2H_2O$
Tsang [74]	H	P		P	P		K			
Massaquoi and Riggs [77,78]	H	P		P	P		E	I	I	
Park and Edgar [79]	H	P	D	P	P			I	I	
Perkins and Sahajwalla [75,80]	H	P		P	P	P	E			I

D = Diffusion limited, E = Equilibrium, H = Heat transfer limited, I = Infinite rate, K = Kinetic (power law) and bulk diffusion, P = Power law kinetics.

As stated earlier, Tsang [74] was the pioneer of the approach and developed a 1D unsteady UCG model considering the two regions of a coal block: the wet zone, and dry zone. In the wet zone, heat conduction and liquid transfer are solved. It was assumed that all drying happens in one moving point (Stefan model). In this model, the movement of the drying front was described using the following equation:

$$\varphi \rho_w (\Delta H_v) \frac{\partial \bar{x}}{\partial t} = \bar{K}_d \frac{\partial T_d}{\partial x} - K_w \frac{\partial T_w}{\partial x}, \ T = T_{vap} \text{ at } x = \bar{x} \tag{34}$$

where ΔH_v is the heat of the evaporation of water, $\bar{x}$ is the location of the drying front.

This model neglected the effects of steam convection and assumed that the drying process was dominated by conduction. In the wet zone, the liquid diffusion was considered as a function of saturation; however, for the dry zone, multicomponent diffusion was calculated using viscous flow. In the dry region, the pseudo-steady-state assumption was adopted to solve mass balance, heat transfer, and gas flux. Pyrolysis reactions were modeled as simultaneous independent reactions with kinetic parameters obtained from the works of Campbell [49]. Porosity and permeability were set to change with the extent of the reactions with correlations obtained from oil well acidization. This model neglected the combustion and provided the heat by setting a high temperature at the boundary. The experimental results of Forrester [81] were used for model validation. The calculated drying front locations with time were found to be in good agreement with the experimental data (Figure 38). However, the effect of the heating rate on the location of the drying front shows that a higher surface heating rate causes a shallower penetration of the drying front into the coal block (Figure 39). This observation indicates the possibility of the coincidence of the drying front with the combustion front at a very high heating rate. Finally, it was concluded from the model that drying cannot be neglected in the UCG model as it is a factor that dominates the heat transfer rate.

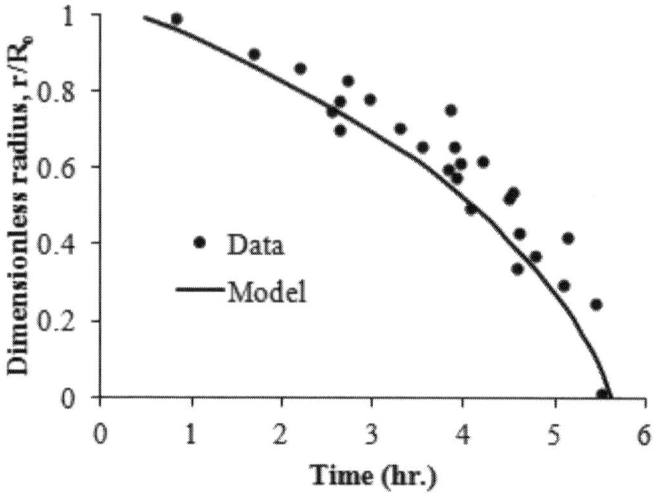

Figure 38. The location of the drying front as a function of time [74].

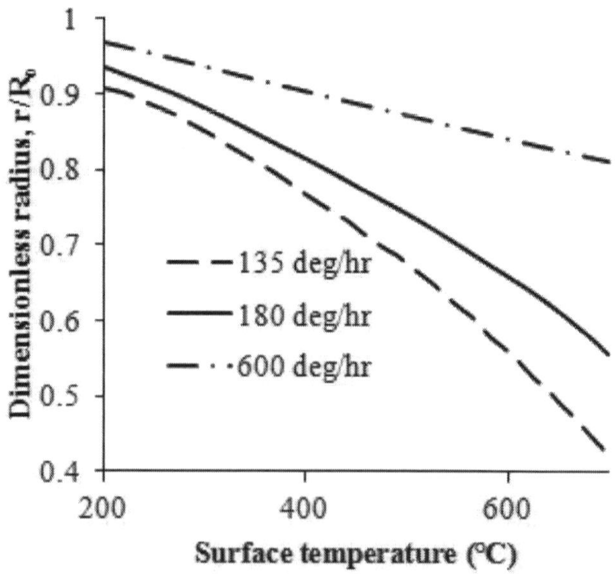

Figure 39. Effect of the heating rate on the location of the drying front [74].

Massaquoi and Riggs [77,78] extended Tsang's 1D model by including the combustion of char and volatiles at the boundary while solving for a steady-state case. Beside the zones shown in Figure 36, they included another zone of bulk gas that is composed of water vapor, CO_2, O_2, and some inert gases outside the gas film. The developed model was used to describe the simultaneous combustion and drying of a wet Texas lignite coal. The flow of gases in the porous dry coal was modeled by Darcy's law and heat transfer was considered by both conduction and convection. However, in the wet zone, only conduction was considered for heat transfer. Radiation heat transfer was also considered between the edge of the ash and the bulk gas. Because of the drop in the total pressure across the dry coal and the high permeability of pyrolyzed coal, mass flux was assumed to happen mainly by viscous flow. Multi-component diffusion was assumed to occur at different zones. They considered almost all the reactions listed in Table 1 except for methanation. Following the approach of Tsang [74], pyrolysis reactions were modeled as simultaneous independent reactions with kinetic parameters obtained from the works of Campbell [49]. Because of steady-state assumptions, all the governing equations were written in ordinary differential equations in terms of the moving coordinates. The cavity wall and the drying front movements were considered to have same velocity. In their model, two different cases for combustion were assumed: (i) homogeneous char combustion with the flame located at the ash; and (ii) heterogeneous char combustion with the flame located at the coal surface, and finally a criterion was defined for the transition from heterogeneous to homogenous combustion. For the case of Wyodak subbituminous coal, their result indicated that the criteria principally depend on the conditions of bulk oxygen concentration, bulk gas temperature, coal surface temperature, and coal burning rate. However, different solutions can be obtained at different conditions. For example, as seen in Figure 40, it is found that, for a given bulk gas temperature, there is a limiting bulk oxygen concentration and an oxygen concentration above the limiting bulk; for a fixed oxygen concentration and bulk gas temperature there are two possible solutions (Figure 40): one with the flame at the coal face (at a lower surface temperature) and another with the flame in the ash (at a higher surface temperature). However, heterogeneous combustion was considered unstable due to the decrease of the coal face temperature with the increasing bulk gas temperature for a constant oxygen concentration. In contrast, homogenous combustion was considered stable because of the experimental observation reported in the literature [82,83,84]. They also indicated a maximum value of the bulk gas temperature above which the flame is always away from the coal surface.

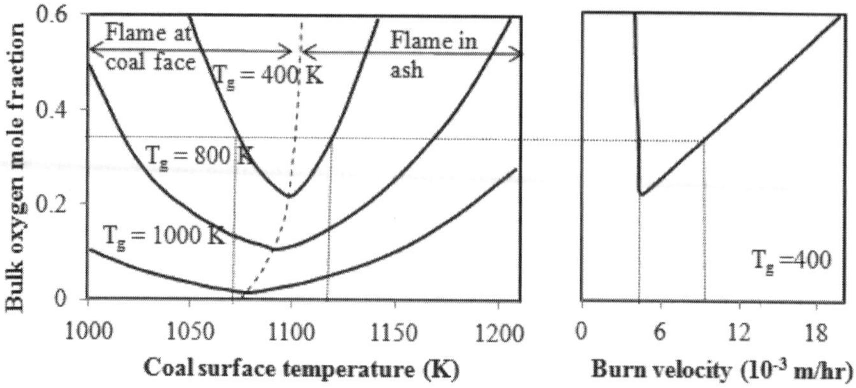

Figure 40. Effect of bulk gas temperature of the combustion of wet Wyodak coal [78].

In addition, they have concluded that the burn rate would be nearly linear with the increase in the concentration of oxygen when the flame front is located in the char face, and the burn rate increases when the flame is located in the gas film (Figure 40). In their model, the effect of the ash layer thickness, coal moisture content, and film transport coefficients were also discussed. A higher film mass transfer coefficient and a lower moisture content and ash layer are observed to shift the transition line towards higher coal surface temperatures corresponding to the higher burning rate for a constant bulk oxygen concentration. For the homogeneous combustion model, the performance of the model was in good agreement with the experimental data [83,84].

Park and Edgar [79] developed an unsteady 1D model with a moving burning front based on the work of Massaquoi and Riggs [77,78] to describe lateral cavity growth in UCG. However, unlike the assumption of Massaquoi and Riggs [77,78], they did not consider having the same velocity for the cavity wall and drying front during the early stage. Beside char gasification, they included coal shrinkage due to drying and pyrolysis, as well as the "self-gasification" of drying and pyrolyzed products (steam and CO_2) in the region between the cavity wall and the drying front in order to consider the additional movement of the cavity wall. The cavity wall movement was determined by the rate of the removal of coal by chemical reactions and the physical movement of the cavity wall due to the shrinkage of the coal. In their simulation, an increase of the cavity growth was noticed during the transient period due to the shrinkage of the coal; however, the cavity growth with and without considering the shrinkage eventually merged into one rate when the steady state was reached (Figure 41). However, they suggested that the movement of the cavity wall due to shrinkage is only important in lab-scale processes and can be neglected in larger-scale processes because of the long timescales in which the transient period is negligible compared to the steady-state period. Their results indicate that cavity growth is controlled by the rate of oxygen transfer to the cavity wall, when the flame is located at the char surface. If the flame is separated from the cavity wall, CO_2 and steam gasification determine the cavity growth rate. The moisture content in coal was found to be a complex parameter for cavity growth. Although the surface temperature decreased with increasing moisture content, the higher moisture content indicated a higher cavity growth at a high coal surface temperature (above 1000 °C). This observation implies that the availability of water in the coal, rather than the reaction temperature, controls the steam gasification reaction rate. They validated their results with the experimental data of lateral cavity growth from a Texas lignite core combustion performed by Poon *et al.* [85]. From the models discussed above [77,78,79], it can be speculated that under oxidizing conditions, the cavity growth rates are controlled by mass transfer.

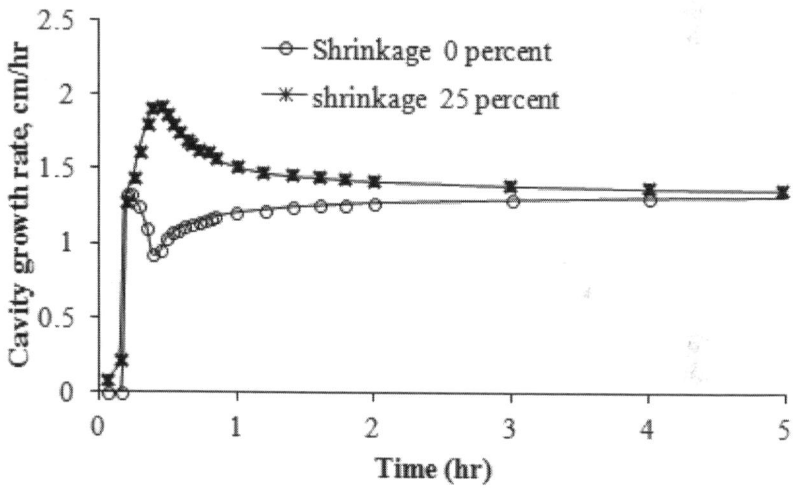

Figure 41. Effect of the shrinkage on the cavity growth rate [79].

Perkins and Sahajwalla [76,80] also developed a 1D transient coal block model with an extension of Tsang's study [74] by including multicomponent diffusion and the random pore model to account for changes of heterogeneous reaction rates with conversion. For multicomponent diffusion, Stefan-Maxwell equations and the bi-dispersed dusty gas model were used for the gas film and dried coal matrix, respectively. Because of experimental evidence (Figure 42) of different sizes of pore distribution in the porous structure of coal, a bimodal porous structure (macropores and micropores) was considered where the macropores act as the major conduits for species flow and the micropores act as the dominant region for gasification reactions.

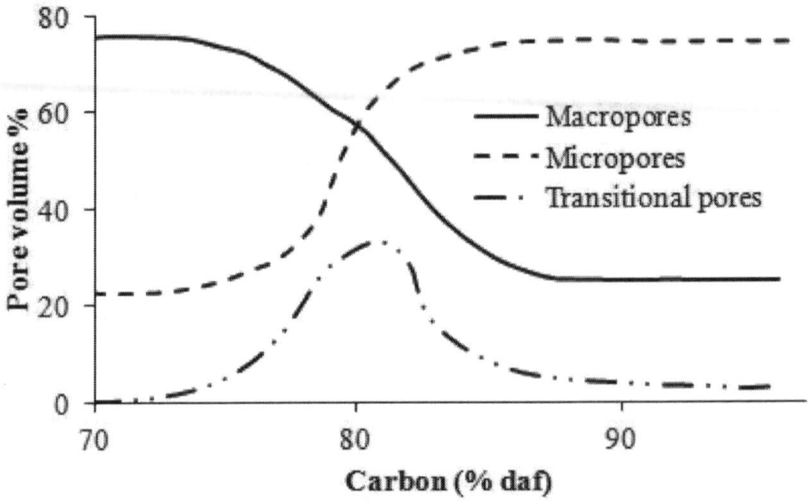

Figure 42. Pore structure of raw American coals on dry and ash-free basis [70].

They have proposed that cavity growth occurs at reducing conditions, therefore only heterogeneous gasification reactions are solved, and required heat was provided by defining a constant temperature at the char surface. The char surface was allowed to participate in radiation heat exchange with its surroundings. For pyrolysis, they followed the work of Tsang [74]. The movements of the drying front and char front were assumed to be equal under pseudo-steady-state conditions. They assumed that solid and gas phases are in thermal equilibrium and bulk gas has a fixed composition that is representative of the product gas. In their first paper [80], they validated their model with the experimental data obtained by Forrester [81]. The drying front location as a function of time was in good agreement after using a scaling factor of 1.2 for thermal conductivity (Figure 43). The temperature profile through the block at different surface temperatures also agreed well with the experimental data. However, when the cavity growth rates as a function of gas temperature for two different bulk gas compositions were compared, no significant difference was observed (Figure 44).

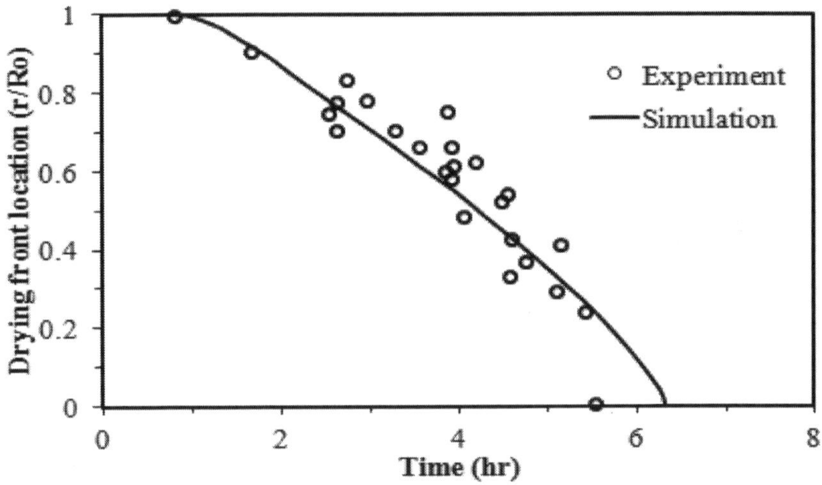

Figure 43. Drying front location as a function of time for cylindrical coal block heating experiments [80].

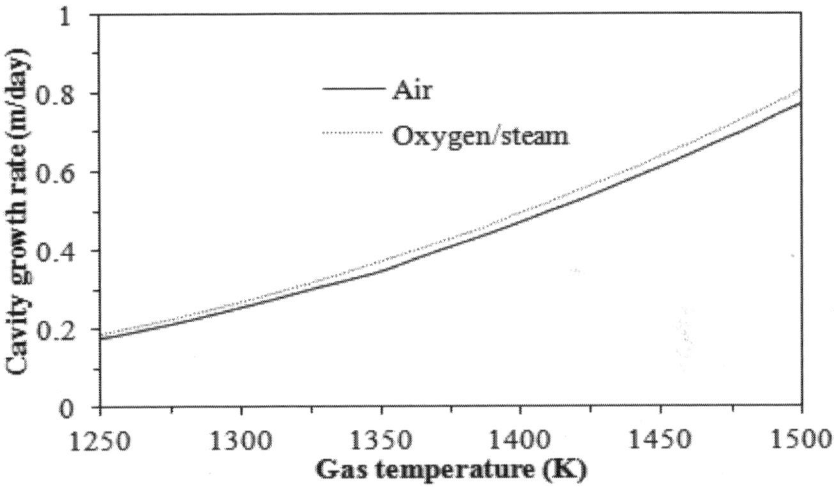

Figure 44. Cavity growth rate, as a function of gas temperature, for two different bulk gas compositions [80].

In a later paper [76], they extended their work by including the ash layer and water influx and investigated the effect of various parameters and operating conditions (e.g., temperature, pressure, water influx, gas composition) on the process performance and cavity growth rate. In order to compare the relative effect on cavity growth due to operating conditions and coal properties, they explored the sensitivity of the cavity growth rate (v_{cv}) to changes in parameters (p_r) using the following finite difference formula:

$$s\left(p_r\right)=\frac{v_{cv}\left(p_{r^o}+\Delta p_r\right)-v_{cv}\left(p_{r^o}-\Delta p_r\right)}{2v_{cv}\left(p_{r^o}\right)}\frac{p_{r^o}}{\Delta p_r} \quad (35)$$

where $s(p_r)$ is the sensitivity of the parameter, $p_r°$ is the base value of the parameter, and Δp_r is a perturbation to the value of the parameter.

A ranking was established according to the sensitivity of the parameters to the cavity growth rate. A positive value was considered as an increase in v_{cv}, while a negative value was considered in reverse order. Table 18 and Table 19show the sensitivity of the cavity growth rate to perturbation in operating conditions and coal properties around the base values, $p°$. Since the summation of the composition of proximate analysis is equal to one, changing any one of these parameters was followed by adjusting the other parameters in order to maintain the constraint of the summation of one.

Table 18. Sensitivity of the cavity growth rate to changes in operating conditions [76].

Parameter	p°	$\Delta p/p^{\circ}=10^{-3}$	$\Delta p/p^{\circ}=10^{-2}$	Rank
Bulk gas temperature, K	1.5×10^{3}	5.276	5.276	1
Water mass flux, $kg/m^2\cdot s$	1.0×10^{-3}	−0.114	−0.115	2
Mole fraction of steam in bulk gas	2.0×10^{-1}	0.066	0.071	3
Ambient pressure, Pa	1.5×10^{6}	0.079	0.067	4
Mole fraction of CO_2 in bulk gas	1.2×10^{-1}	−0.009	−0.009	5
Mole fraction of H_2 in bulk gas	1.6×10^{-1}	0.008	0.008	6
Heating rate, K/s	1.0×10^{-2}	0.006	0.007	7
Mole fraction of CO in bulk gas	8.0×10^{-2}	0.005	0.005	8

Results of their parametric study indicate that temperature of the coal surface, water influx, and coal composition have the highest impact on the cavity growth rate. Decreasing fixed carbon was found to increase the cavity growth rate. This implies that the cavity growth rate is more for low-rank coal due to the smaller fraction of fixed carbon in the proximate analysis. Perkins and Sahajwalla [76] graphically demonstrated the variation of cavity growth due to the variation of different parameters through their modeling. In addition, they incorporated the coal spalling

Table 19. Sensitivity of the cavity growth rate to changes in coal properties [76].

Parameter	p°	$\Delta p/p^{\circ} = 10^{-3}$	$\Delta p/p^{\circ} = 10^{-2}$	Rank
Critical conversion of carbon at which the coal spalls	9.50×10^{-1}	−3.719	−3.741	1
Mass fraction of fixed carbon from proximate analysis	3.35×10^{-1}	−1.359	−1.370	2
Solid material density (kg/m^3)	1.47×10^{3}	−0.913	−0.925	3
Mass fraction of ash from proximate analysis	2.40×10^{-1}	0.714	0.728	4
Ash layer thickness (m)	5.00×10^{-3}	−0.689	−0.703	5
Mass fraction of volatile matter from proximate analysis	3.00×10^{-1}	0.589	0.576	6
Mass fraction of moisture from proximate analysis	1.25×10^{-1}	0.344	0.331	7
Emissivity of ash (for radiation)	8.50×10^{-1}	0.114	0.127	8
Initial specific surface area (m^2/m^3)	1.23×10^{7}	0.105	0.105	9

model by changing the value of the critical conversion, $X_{critical}$. Due to lack of information on coal spalling, they considered the critical conversion of fixed carbon as an indicator of coal spalling by assuming a level of carbon conversion at which some fixed carbon and volatile matter are unconverted and detach from the coal block due to mechanical stress and breakage. Critical conversion was found sensitive to the bulk gas temperature, molar flux of product gas, and water influx.

Finally, they compared the estimated cavity growth rate from their simulation with six different field trials (Hanna I, Pricetown, Large Block Test No. 5 (LBK-5), Partial Seam CRIP (PSC), Rocky Mountain I (RM I), and El Tremedal) using the available operating conditions and coal properties of those field trials and obtained a good agreement with the data from the field trials (Figure 45) despite some uncertainty of the field data. Although they conducted a detailed study of the coal block model, they did not consider char combustion and other gas phase reactions (except the water-gas shift reaction) which are strong functions of the temperature. Rather, the assumption of a constant temperature and fixed gas composition at the boundary limits the applicability of the model to field trials.

The special feature of the coal slab model is the tracking of the drying and combustion front movement. This model can successfully demonstrate the drying and devolatilization behaviour of large coal particles; however, this model is yet to be validated using UCG field trial data. Due to the assumption of semi-infinite coal slab, it can be speculate that this model is only good for a thick coal layer. All the models developed thus far by considering the coal slab are one-dimensional (1D). As a result, information on cavity shape cannot be obtained from these models.

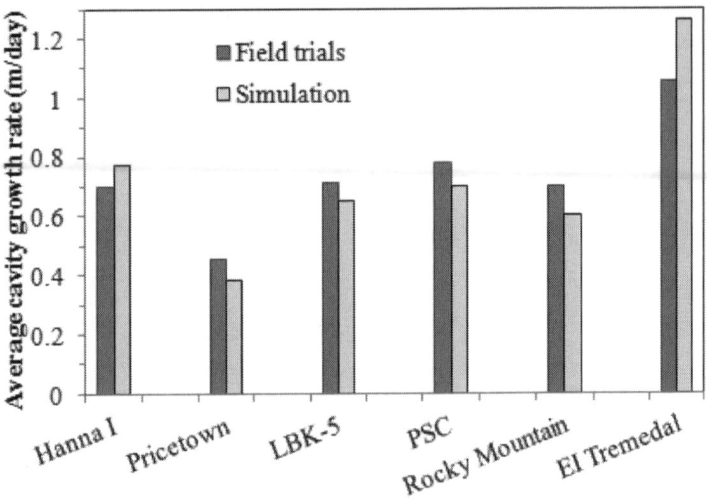

Figure 45. Comparison of estimated cavity growth rates from six UCG field trials with model simulations [80].

2.4. Other Approaches for UCG Modeling

2.4.1. Reactor Models

Over the years, there have been a few studies in which the UCG process is described as a series of ideal reactors. Such models can be used to predict product gas composition without considering complex phenomena occurring underground.

Chang *et al.* [86] used this approach to predict the resource recovery during the UCG process. In this model, the domain was divided into cylindrical elements where each element consists of two distinct zones: rubble and void. It was assumed that the rubble zone acts as an ideal plug flow reactor (PFR), while the void space was assumed to be perfectly mixed as a continuous stirred tank reactor (CSTR). Each element was allowed to change based on coal consumption. The fraction of flow in each zone was changed to match the results of the field trials. The spalling and water influx were calculated based on empirical correlations. Their modeling results highlight the importance of turbulence in field trials. Therefore, they concluded that small laboratory-scale experiments could not be scaled to field trials.

In order to obtain *in situ* measurements, the stimulus response technique (tracer injection) was used in some field trials to study the flow distribution inside the underground cavity. Some researchers have used the response of tracer material from the experiment to characterize different zones in the underground cavity and track the cavity growth.

Based on the residence time of tracer elements, the cavity is modeled as an arrangement of ideal reactors.

Debelle *et al.* [87] used various chemical engineering reaction models to fit the quasi-exponential decay of the tracer response from a UCG experiment in Thulin, Belgium. In the experiment, ^{133}Xe was selected as a tracer because of its detectability from a complex gaseous mixture by using a γ-ray detector with high precision. The quasi-exponential decay of the tracer was considered as an indication of mass exchange between a flowing fluid and a dead zone. That is why the models were developed to simulate the flow conditions based on the material exchange between flowing fluid and the dead zone. The film theory was applied to simulate the mass transfer between a stagnant volume fraction and the following fluid volume fraction. However, for the interaction between the flowing fluid and porous zone, molecular diffusion was considered. The dead zone was represented either by a stagnant zone or by a homogenous porous zone where materials exchange with the flowing fluid. Various stages of the process such as reverse combustion linking and forward gasification were represented using regression models. For the gasification period, it was concluded that the cavity offers a high level of back-mixing, resembling a well-mixed reactor with a small dead zone.

Thorsness [88] performed tracer experiments in a UCG field test at the Hoe Creek trial III. Their goal was to investigate the extent of the water influx and steam injection requirements. The tracer response curve was regressed with a series of CSTRs in two parallel paths.

Pirard *et al.* [89] considered the El Tremedal trial in Spain where helium was used as a tracer. In their study, two classical engineering models were used: (i) the combined model composed of a succession of stirred tanks in a series exchanging matter with an adjacent porous zone (the STEP model) and (ii) the combined model composed of a succession of stirred tanks in a series exchanging matter with a stagnant zone (STES model). Both models are three-parameter models. The parameters are: the volume fraction occupied by the flowing fluid, the number of tanks in the series, and the dimensionless parameter characteristic of the pore diffusion. By these methods, they were also able to detect cracks or breaks in the coal seam. Their results indicated the STEP model as the best fitted model, and finally they concluded that the cavity behaves almost like a small number of CSTRs in a series with a high level of back-mixing.

Recently, Daggupati *et al.* [90] used computational fluid dynamics (CFD) to describe the cavity as series of ideal reactors (Figure 46). The residence time distribution (RTD) of gases was calculated based on CFD results. This approach was validated by laboratory-scale tracer experiments. Based on RTD and velocity distribution, the cavity was simulated as a network of ideal reactors. The proposed arrangement of reactors consists of a series of five CSTRs and a side stream of five small CSTRs, followed by a plug flow reactor (PFR). They concluded that as the cavity size increases, the cavity behaviour changes from PFR to CSTR. They have implemented their reactor model in ASPEN PLUS (chemical process optimization software) and compared the results with CFD predictions.

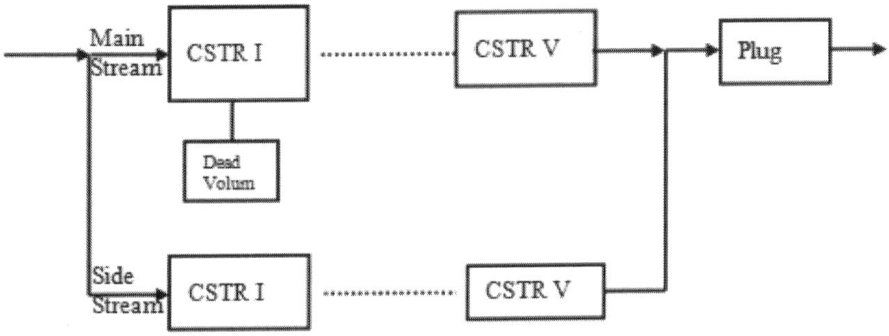

Figure 46. Proposed reactor arrangement for the cavity in the works of Daggupati *et al.* [90].

2.4.2. Probabilistic Simulation

In a unique approach to model the UCG process, Batenburg and Bruining [91] adapted a probabilistic simulation to describe the forward gasification phase in a 2D UCG geometry (Figure 47). The model used stream function on the grid blocks. The derivative of the stream function along the ash-char interface was interpreted as the possibility that the interface was moving in a certain block. This model is very limited, as the temperature of the gases and composition are constant in each region.

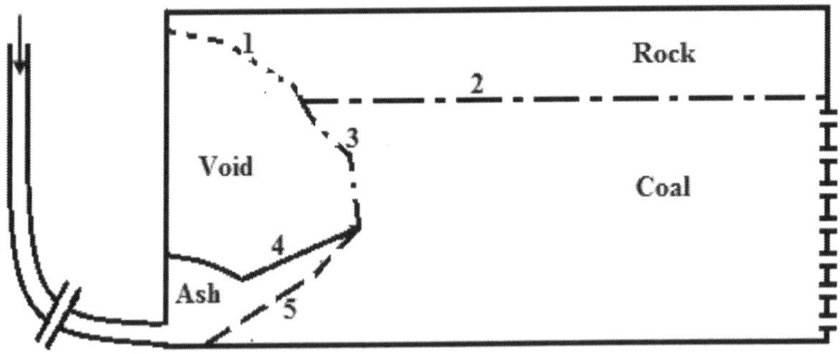

Figure 47. Probabilistic simulation of the UCG process by possibility of movement of the following interfaces: 1.void-rock; 2. coal-rock; 3. void-coal; 4.void-ash; 5. ash-void [92].

From the same research group, Biezen *et al.* [93] extended the concept of the probabilistic simulation by including the movement of several interfaces including ash-void, ash-coal, void-coal, and void-rock. The model consists of two modules: one module solves flow in the entire domain and the other selects a block of coal/rock for gasification or roof spalling. However, in order to consider roof spalling, they incorporated a very simple spalling model which assumes a constant spall rate. A single streamline was selected randomly and based on the interface located on the streamline, and relevant physical phenomena proceeds. For example, in the void-coal interface surface gasification reactions occur and the corresponding amount of ash accumulates at the bottom of the cavity. They later extended their model to simulate cavity evolution in a 3D domain [92]. This model was intended to simulate the development of a gasifier from early times (gasification in a permeable bed) to a developed stage (gasification in a channel). They did not solve the mass and energy balance together. As a result, the model is limited for making predictions of performance. However, the model is more capable of interpreting previous field trials. The application of this model is limited due to its complexity and unphysical assumptions such as constant temperature at each domain.

2.4.3. CAVSIM (a computer code for 3D, axisymmetric UCG model) Model

All of these reviewed modeling efforts so far have the deficiency of focusing on some aspect of the UCG process while neglecting others. During UCG trials in the US, LLNL funded many modeling studies for the UCG process. Several 1D models were developed and validated with certain series of data from field trials. These segregated models were later combined together to form the CAVSIM process model that represents 15 years of continuous UCG research and development in the US [23]. The model is applicable for predicting lateral cavity growth of thick and shrinking coal seams in which oxygen is injected at the bottom of the coal seam. The model was based on some major assumptions such as: the cavity was axisymmetric around injection; thermal radiation was the main heat transfer mechanism in a well-mixed void space; cavity growth was dominated by the thermo-mechanical failure of the wall, and a packed bed of char and rubble forms over a thin layer of ash. Various sub-models were incorporated into this model to quantify major phenomena in UCG such as water influx, flow dispersion through a rubble bed, sidewall growth of the cavity due to reactions, and the spalling of the cavity roof into the domain. This model reproduced results of some of the field trials in the US by tuning several parameters that were introduced in the model. Therefore, the model could not be used for predicting UCG behaviour in other seams, as these parameters are not known prior to the process.

2.5. CFD Software Tools

From the above reviews, it is evident that the understanding of gasification for a particular coal seam in a specified location with its boundary conditions and various time-varying phenomena are imperative in order to identify the appropriate UCG model. Although several models with varying levels of complexity have been published over the past years, the applicability of these models is limited to specific and isolated cases and the models could hardly be used to predict the performance of UCG trials. Furthermore, most of these models are in 1D or 2D, while the field trials reveal a 3D non-regular cavity shape. Thus, it is of the utmost importance to develop a 3D model in order to predict the performance of UCG trials. The earlier models compromised between model complexity and simplicity in order to get faster computation runtimes. However, currently, with the advent and advancement of different modeling software tools and powerful computers, it is possible to develop a 3D model with a varying limit of complexities and a much faster runtime. There are several commercial CFD software tools which were used successfully for the UCG models. Some of the popular tools are:

- Ansys Fluent.
- CMG Stars.
- Star CCM+.
- Comsol Multiphysics
- ABAQUS

The basic difference of the above tools is the discretization approaches. Ansys Fluent and Star CCM+ are finite volume-based solvers whereas CMG Stars is a finite difference-based solver. Comsol Multiphysics and ABAQUS follow the finite element approach. Recently, a number of investigators used CFD software tools in order to develop the UCG models. Table 20 shows a list of the researchers who developed a UCG model using different software tools.

The commercial CFD software packages are intended for general use. As a result, not all the functions required for the development of UCG models are available in those packages. However, most of the packages have the facility to incorporate user-defined functions. Investigators usually take this benefit and develop their own user-defined functions for their model in order to overcome this shortcoming of the software packages.

Table 20. List of the researchers who used CFD software tools in order to develop their UCG model, recently.

CFD Software Tools	Investigators	
	3D Model	2D Model
Ansys Fluent	Shirazi *et al.* [50]	Perkins and Sahajwalla [61] Luo *et al.* [58]
CMG Stars	Kariznovi *et al.* [9] Nourozieh *et al.* [12] Seifi *et al.* [94]	
Star CCM	Nitao *et al.* [95]	
Comsol Multiphysics		Shirazi [27] Srikantiah *et al.* [96]
ABAQUS	Yang *et al.* [97] Sheng [98]	Sarhosis *et al.* [99]

2.5.1. 3D Models Using CFD Software Package

There are few 3D models developed so far in the context of UCG. As mentioned earlier, Biezen [92] developed a 3D model for cavity growth in 1995 that appears to be the first 3D UCG model. However, his model was limited due to the absence of any heat transfer calculation and the assumption of a constant reactor temperature. Recently, due to the rapid development of commercial simulation tools for handling complex transport phenomena and chemical kinetics, the development of a 3D UCG model using commercial software tools is especially noticeable. Nouroziehet *al.* [12] developed a 3D model for CRIP configuration using CMG Stars, a reservoir simulation software, and discretized the domain into coarse cubic meshes of 25–50 cm, typical to reservoir solvers. Their computational domain was similar to the proposed field trials in Alberta (15 m × 5 m × 9 m). They have introduced an arbitrary clay layer in the seam to investigate the effect of inert layers on the process; however, thermo-mechanical failure of this layer is not included in the model. They included chemical reactions (all the reactions listed in Table 1 except reaction 9) and the conduction/convection of heat and species in their model. They did not make any attempt to treat heterogeneous reactions separately. Except for the methanation reaction, all other reactions were modeled using the power law and the temperature-dependent rates were described by the Arrhenius correlation. Their results show a large degree of grid dependence. Resizing grids affected the temperature distribution and altered the gas composition at the producers (Figure 48 and Table 21).

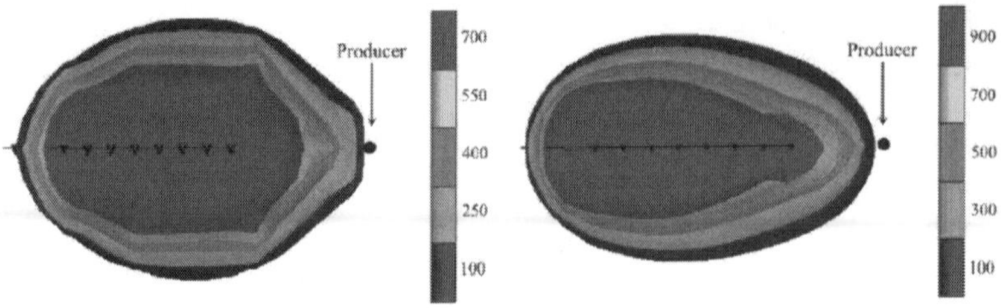

Figure 48. Effect of the grid size on the temperature distribution: large grids (left) and small grids (right) (reproduced from Nourozieh *et al.* [12]).

Table 21. Effect of the grid size on product gas composition [12].

Gas Composition (Mole Fraction)		CO_2	H_2	CO	CH_4
Grid size	0.5 × 0.5 × 0.5 m	0.3490	0.2887	0.2206	0.1416
	0.25 × 0.25 × 0.25 m	0.2847	0.2557	0.2864	0.1732

Because of the 3D model, they were able to exhibit the cavity growth and shape (combustion front) that was nearly the shape of tear drops (Figure 49). Their study also showed an increase of gas temperature with an increasing oxygen flow rate (keeping the H_2O/O_2 ratio constant) and decreasing H_2O/O_2 ratio (keeping the O_2 flow rate constant). With the increasing pressure and H_2O/O_2 ratio, a significant increase of methane and decrease of hydrogen is noticed in their model prediction. This phenomenon was explained as the favourable condition for methanation at high pressure. However, the methane concentration was found higher even in low pressure conditions as compared to other studies. As they did not consider reaction 9, which converted H_2 to H_2O and reduced the concentration of H_2, the presence of a higher H_2 concentration possibly favoured the methanation reaction. They did not present any validation for their model. They have extended their modeling to simulate CRIP with successive ignition points [94]. Again, no validation or comparison with experimental data was reported.

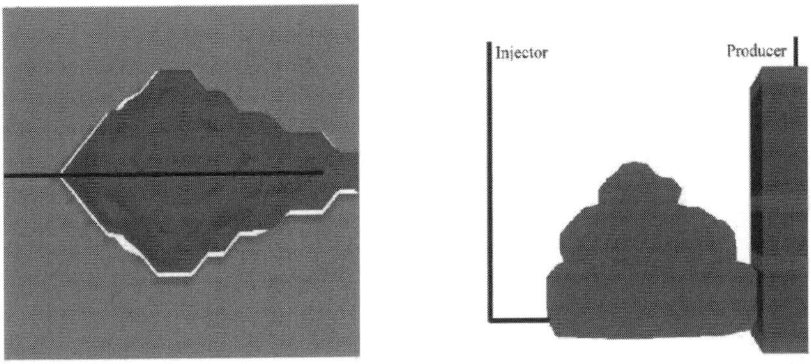

Figure 49. Cavity shape corresponding to the geological structure after 10 days of simulation: x-z cross (right) and x-y cross (left) (reproduced from Nourozieh *et al.* [12]).

Recently, Shirazi *et al.* [50] developed a small-scale 3D packed bed model (3 cm × 1.5 cm × 2 cm) for a shrinking coal seam using Ansys Fluent. The following set of governing equations is solved in his model:

Continuity equation:

$$\frac{\partial}{\partial t}\left(\varphi\rho_g\right) + \nabla\left(\mu\rho_g\bar{v}\right) = S_{\varnothing} \tag{36}$$

Momentum equation:

$$\frac{\partial}{\partial t}\left(\varphi\rho_g\bar{v}\right) + \nabla.\left(\varphi\rho_g\bar{v}\,\bar{v}\right) = \mu\nabla^2\varphi\bar{v} - \varphi\nabla p + \varphi\rho_g\bar{g} - \left(\frac{\varphi^2\mu}{\alpha}\bar{v} + \frac{\varphi^3 C_2}{2}\rho\left|v\right|\bar{v}\right) \tag{37}$$

The flow in porous medium is implemented in Ansys Fluent by adding a source term into the original Navier-Stokes equation.

Energy equation:

$$\frac{\partial}{\partial t}\left(\rho_g C_{p_g}\varphi + \rho_s C_{p_s}\left(1-\varphi\right)\right) + \rho_g C_{p_g}\left(\bar{v}.\nabla T\right) = \nabla.\left[k_{eff}\nabla T\right] + \Sigma rate_i \Delta H_i \tag{38}$$

In the model, it was assumed that gas and solids that form the porous media are in thermal equilibrium in each cell; thus, one energy equation was used to describe heat transport in each cell. Only heat conduction was considered.

Transport of species:

$$\frac{\partial}{\partial t}\left(\rho_g Y_i \varphi\right) + \rho_g\left(\overline{v}.\nabla Y_i\right) = \nabla.\left[\rho_g D_i \nabla Y_i\right] + \Sigma rate_i \tag{39}$$

Transfer of species was described by the convection-diffusion equation in porous media.

The cavity development is tracked by the decrease of porosity due to chemical reactions and thermal effects. Porosity increases as the solids are being consumed by reactions or species transfer from the solid phase to gas phase due to thermal effects. The change of porosity with chemical reactions is calculated with the following equation:

$$\frac{\partial}{\partial t}\left(\varphi\right) = S_\varphi \tag{40}$$

where S_φ is the source term for porosity based on reactions in the coal.

Coal block is considered to be a porous medium with a defined initial porosity and permeability. However, the porous media model in Fluent does not allow for solids to participate in reactions. That is why no separated mass conservation equation was solved for the solid phase. Therefore, in order to simulate and track the solid and gas/solid reactions, a virtual solid phase was defined and a number of user-defined scalars (UDS) were written to be implemented in Fluent. Three scalar equations calculate the consumption of moisture, volatile matter, and fixed carbon following a general form of UDS equations in Fluent:

$$\frac{\partial}{\partial t}\left(\rho_g \varnothing \varphi\right) + \frac{\partial}{\partial x_i}\left(\rho_g v_i \varnothing - \frac{\partial \varnothing}{\partial x_i}\right) = S_\varnothing \tag{41}$$

The ash content of the coal is assumed to segregate after consumption of the coal and no ash layer remains over the coal. This assumption has been verified by several experiments and field trials [34,68]. In the model, the changes in porosity and permeability with time due to the drying, pyrolysis, and reactions with coal were included. The gas properties are taken from the Fluent database. Heat capacities of gases were taken following the precise polynomial model.

Their model revealed the effect of different parameters on cavity growth as well as cavity shape and size. Lower oxygen flow rates at the inlet led to a more bulbous cavity shape, whereas at higher flow rates, the cavity became elongated in the direction of the production well. In addition, increasing the inlet oxygen flow beyond a certain value decreased the concentration of CO and H_2 in syngas due to the excessive combustion of these gases. Initial permeability of the seam also had a strong influence on the growth rate and shape of the cavity. The higher coal permeability led to faster growth of the cavity and resulted in a much wider cavity (Figure 50).

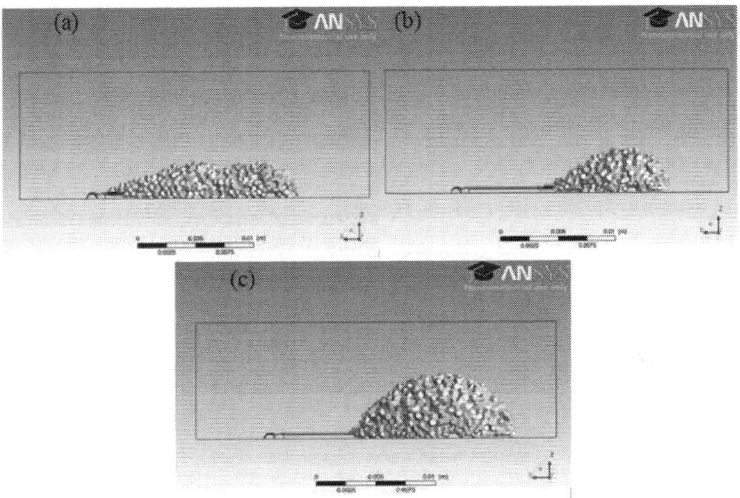

Figure 50. Cavity shape for cases with different initial permeabilities: (**a**) 1 mD, (**b**) 100 mD, and (**c**) 10 D (reproduced from Shirazi *et al.* [50]).

Unlike some model predictions (*i.e.*, Daggupati *et al.* [90]), a very negligent effect of the temperature of the inlet oxygen was observed on cavity growth. Their model was validated with the lab-scale experiments of Daggupati *et al.* [34] in terms of cavity growth (Table 22). The agreement between the model calculations and the lab experiments indicates that the relevant physics and reactions of their model for small coal seam geometry are useful in the design of the gasification process.

Table 22. Comparison of the model-predicted cavity dimension with the experiments [47].

Dimension	Model [42]	Experiments [25]
Forward length	1	1
Backward length	0.429	0.433
Height	0.399	0.517
Width	0.404	0.425

2.6. Sub Model: Drying and Pyrolysis

The drying and devolatilization are the integral parts of the UCG process. During UCG, drying, pyrolysis, and combustion of the coal occur simultaneously [77,78]. The drying is related to moisture release whereas devolatilization is related to the production of char by the evolution of volatile matters and tars. The development of drying and devolatilization sub-model is also as important as the development of the cavity growth model as all these processes are involved in UCG cavity expansion.

2.6.1. Drying Sub-Model

Drying is usually considered as the release of the unbound moisture in the coal at the evaporation temperature of water. Moisture content is an integral part of the coal seam. However, the percentage of moisture present in coal depends on coal ranks and sources. Although drying was considered with importance in most of the packed bed and coal slab models, this phenomenon has been neglected in most of the channel models. However, instead of a drying rate, the drying front movement calculation is found to be the main focus for coal slab models.

In order to interpret drying phenomena, the following models are well recognized [74]:

- The diffusion theory,
- The capillary flow theory,
- The evaporation-condensation mechanism,
- Simultaneous energy and moisture transfer in porous media

The evaporation-condensation mechanism is generally found to be used for wet coal where mobile water is available or in a situation where water influx is considered [47,48,100]. In construct, drying is usually considered for the evaporation of moisture content (fixed water) in the coal. This moisture can be present in the coal's structure as unbound, weakly bound, and chemically bound. The chemically bound moisture is released at higher temperatures around 200–400 °C [101], while the unbound moisture is released at the evaporation temperature of water [102] which is considered as the standard drying temperature for most of the UCG models. Some UCG models assume that drying occurs at a front layer in which the temperature is kept constant, equal to the water evaporation temperature, and drying is assumed to be governed by an Arrhenius-type reaction expression:

$$-\frac{dM}{dt} = -rate_{drying} = k(M^* - M) \quad (42)$$

where M is the moisture content of coal at any moment and M^* is the moisture content of coal from proximate analysis.

However, the release of weakly bound and chemically bound moistures at higher temperatures is not correctly governed in these models. On other hand, some models recognize that the moisture in the coal is distributed throughout the coal and is not located solely at the gas/solid interface. At high pressure, the drying temperature depends on the water vapor pressure. Considering this fact, Thorsness *et al.* [48] developed an expression for the drying rate that is less than the evaporation rate for mobile water. The evaporation/condensation rate was considered by a pure mass transfer controlled expression as follows [48]:

$$-rate_{evaporation/condensation} = k_y\left(\frac{p_w}{P} - y_w\right),\ \text{for } p_w \leq P \quad (43)$$

where k_y is the mass transfer coefficient, y_w is the mole fraction of water in gas, p_w is the vapor pressure of water, and$P$ is the total pressure. The temperature and vapor pressure are given by:

$$p_w = 10^5 exp\left(12.61 - \frac{4690}{T_d}\right) \quad (44)$$

$$T_d = \beta\, T_s + (1 - \beta)\, T_g \quad (45)$$

where β is the weighing coefficient. However, most of the models considered β = 1 which reduced the drying temperature to the surface temperature ($T_d = T_s$). From Equation (44), Perkins and Shajwala [100] deduced an expression for the drying temperature for calculating the drying temperature at high pressure:

$$T_d = \frac{4690}{12.61 - \ln\left(\frac{p_w}{101325}\right)} \quad (46)$$

For the drying rate, Thorsness *et al.* [48] expressed the basic rate for spherical symmetric coal particles as follows:

$$\frac{dM}{dt} - rate_{drying} = k_y \frac{\gamma^2}{3\eta}(\frac{p_w}{P}\frac{\rho_w}{\rho_w^o} - y_w) \quad (47)$$

where $\rho°_w$ is the uniform initial moisture density and parameters γ and η deal with the internal resistance. Parameter $\eta$is determined from the following expressions:

$$\eta = \frac{k_y \rho_w}{2P\rho_w^o D^{eff}} \quad (48)$$

Parameter γ is the first Eigenvalue given by:

$$tan\gamma_n = \frac{\gamma_n}{(1-\eta)} \quad (49)$$

However, for simplicity, Winslow [47] and Thorsness and Rozsa [42] calculated the drying rate from the evaporation rate described by Equation (43) multiplied by a factor less than one (<1). Due to the inherent dependency of the mass transfer coefficient on particle diameters in the Thorsness [48] model, the drying rate expression also closely depends on the initial size and shape of the coal particles.

2.6.2. Pyrolysis Sub-Model

During devolatilization, the releasing of the volatiles occurs in the following order: chemically bound H_2O, CO_2, CO, CH_4, tars, and H_2 [15]. Some other hydrocarbons also evolve at a lower concentration. Due to the high residence time of gas and high pressure in the underground coal gasification process, tar would likely undergo secondary reactions such as cracking, and tar yield would be lower compared to the surface gasifiers [80]. Thorsness *et al.* [48] used an empirical formula C_9H_c for tar and a trace amount of hydrocarbon not accounted for in the volatiles mentioned above. Pyrolysis of coal can be represented by the following reactions:

$$Dried\ Coal \rightarrow Char + Tar + Ash + \text{Volatiles} \quad (50)$$

The volatiles can be further represented by:

$$Volatiels \rightarrow a_1CO_2 + a_2CO + a_3CH_4 + a_4H_2O + a_5H_2 + a_5\ Other\ hydrocarbons \quad (51)$$

The matrix of the coefficient can be obtained from the element balances of the atoms involved in the above equation and the data from the proximate and ultimate analysis of the coal. Some models obtained the stoichiometric coefficients based on the work of Cambell [49] or Suuberg *et al.* [103] for subbituminous coal or lignite coal, respectively.

In order to account for tar, Shirazi *et al.* [42] followed the kinetic model proposed by Boroson and Howard [90], who considered tar cracking with the following overall reaction:

$$tar \rightarrow b_1CO_2 + b_2CO + b_3CH_4 + b_5H_2 \quad (52)$$

The stoichiometric coefficients and kinetic parameters can be found from the work of Boroson and Howard [104].

The rate of gas evolution is different at different temperatures. Figure 51 shows an experimental observation of gas release from coal as a function of temperature. A similar observation was reported by Campbell [49]. It is observed that different species evolve at different peak pyrolysis temperatures.

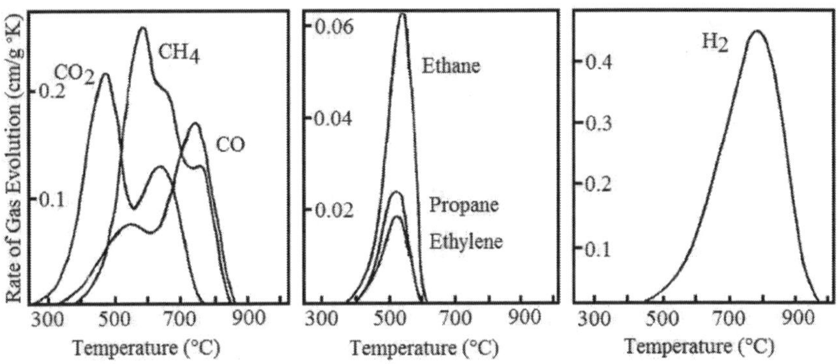

Figure 51. Effect of temperature of gas evolution during the pyrolysis of Wyodak coal at a heating rate of 3.33 °C/min under atmospheric pressure on a dry basis (reproduced from Juntgen and van Heek [105]).

There are multiple steps of decomposition during the pyrolysis process. However, in order to avoid complications, Van Krevelen *et al.* [106] proposed the whole pyrolysis process as a single-step decomposition as follows:

$$-\frac{dVM}{dt} = -rate_{pyrolysis} = k\left(VM^{*} - VM\right) \tag{53}$$

where VM is volatile matter content at any moment and VM^{*} is the effective volatile matter content of coal. VM^{*} is usually assumed to be the same as volatile matter from the proximate analysis. The rate constant k is typically related with the temperature by an Arrhenius expression:

$$k = k_0 \exp(-\frac{E}{RT}) \tag{54}$$

In order to avoid complicacy, some models used the above equation for pyrolysis [9,12,50]. Kariznovi *et al.* [9] summarized the experimental kinetic parameters for different coal. However, the lumping of a number of independent first-order reactions into a single first-order reaction tends to underestimate the frequency factor, k_o, and the activation energy, E. In order to avoid this complexity, Juntgen and Van Heek [105] proposed simultaneous independent reactions with different reaction constants to describe the evolution of different species during pyrolysis as follows:

$$\frac{dVM_i}{dt} = -rate_{pyrolysis} = k_j(VM_j^{*} - VM_j),\ \text{j} = 1, 2 \ldots.\ \text{number of volatiles} \tag{55}$$

$$\text{where, } k_j = k_j \exp(-\frac{E_j}{RT}) \tag{56}$$

Campbell [49] and Suuberg *et al.* [103] carried out excellent work to estimate the frequency factor and activation energy for each species for the pyrolysis of Montana lignite and Wyodak subbituminous coal, respectively. A number of researchers [74,76,77,80] included simultaneous independent reactions for pyrolysis with the quantities of pyrolysis gases evolved and the corresponding kinetic parameters obtained from the works of Suuberg *et al.* [103] or Campbell [49] depending upon the types of coal they used in their models.

Similar to the UCG model, the location of the drying and devolatilization zone depends on the approach of the modeling. Figure 52 shows the reaction zones in the UCG process for different stages of gasification. Usually, a UCG model based on the packed bed approach followed Figure 52a for drying and devolatilization. On the other hand,Figure 52b was followed by the model based on the channel approach.

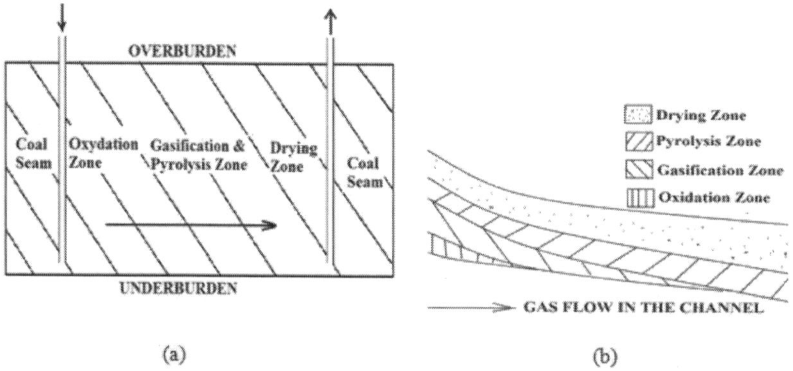

Figure 52. (**a**) Reaction zones along the direction of gas flow; (**b**) Reactions zones surrounding a channel (reproduced from Tsang [74]).

2.7. Consideration of UCG Modeling Parameters

As mentioned earlier, UCG is site-specific. Controlling UCG is less flexible as compared to surface gasifiers because some important variables such as the thickness, quality of the seam, moisture contents, and other physical and chemical parameters are unique properties of the natural location and cannot be changed. Most of the models indicated the sensitivity of the process to the properties of the seam and the change of these properties with temperature. In order to obtain the best possible results from governing equations, appropriate solid and gas properties are required. Most of the models discussed above used either constant properties or transient properties obtained from the approximate model or experimental results of others. However, for a realistic model of any specific UCG site, the physical properties of coal and their changes with temperature and pressure, such as thermal conductivity, specific heat, density, porosity, and permeability, *etc.*, need to be considered. Similarly, properties of gas species and their changes with temperature and pressure, such as density, viscosity, thermal conductivity, diffusivity, and heat capacity, *etc.*, need to be considered also. Some experiments might be required in order to formulate these dependencies for the specific seams. In order to enhance the features of the 3D model and to apply the model in UCG field trials, it is imperative to incorporate the coal properties, the change of properties with temperature, and the coal seam geology of the specific site into the model.

In order to avoid model complications and to reduce computation time, some models assumed that gas and solids in the porous media are in thermal equilibrium and that is why they used one combined heat transfer equation. However, there is a large difference between the solid and gas characteristic timescale. It is assumed that gas reaches the steady state before any change happens in the solid [46]. Thus, two separate heat transfer equations for gas and solid, respectively, seem to be more appropriate to incorporate into the model. In some models, the mass conservation of the solid is avoided since drying, pyrolysis, and chemical reactions are considered in mass conservation of gas. However, in order to account for the mass loss of solid as well as ash production, the solid species balance seems to be reasonable for incorporating into the model. The choice of the appropriate use of reaction rates is always challenging due to the complex situation of UCG operations. As these values are obtained from experiments, a number of experiments were performed in order to obtain reaction rates. However, obtaining a reaction rate for a specific reaction needs to consider the operating pressure and temperature similar to UCG operational conditions.

There is hardly a UCG site without any water influx from the surrounding strata. Thus, in a UCG site, water influx is unavoidable and the gasification is performed above the hydrostatics pressure in order to control the water influx; this allowed the water to be used for steam gasification in order to utilize the heat of combustion and increase the concentration of CO and H_2 in the product gas. It is imperative to include a water influx formulation that is practicable for a specific site's coal hydrology. Although some investigators claimed a steam/oxygen ratio of 2.5 as optimum for UCG [34,107], for different sites and coal seams, it is also necessary to explore the appropriate steam/oxygen ratio.

3. CHALLENGES OF UCG COMMERCIALIZATION

In spite of the significant advantages of UCG technology and repeated trial successes, UCG is still not close to full commercialization. Historical evidence shows that the commercialization of a technology requires a reasonable balance among the technological, environmental, social, and economic factors. In terms of technology, a lot of problems associated with UCG were solved already in the last few decades. The improvement was focused on obtaining (i) good and nearly constant gas quality and heating value; (ii) high thermal efficiency and resource recovery; (iii) high process efficiency and equipment reliability; (iv) high predictability;

(v) good control on the combustion front; (vi) handling bituminous coal; and (vii) handling the depth of the coal and gas clean-up [108]. The improvements also focused on identifying site specificity and a preferable well spacing. However, a few critical issues are still remaining, such as subsidence, ground water depletion and contamination, controlling ground water influx, and gas leakage. These issues associated with environmental risk are considered as the greatest threats for commercialization. Thus, the main challenge is to address the professional management of the environmental risks [108,109].

Several field trials and a few commercial applications of UCG indicate that UCG is very site -specific [10,11]. The technical and commercial risk factors can be decreased considering appropriate site selection by assessing the site, particularly the depth of the coal seam as well as of the overlying and underlying strata, and the geotechnical factors such as the strength, jointing, and deformability of the overlying strata, the surrounding geology, and the hydrogeology [10,109]. In order to proceed with site selection, more detailed knowledge is required for subsidence and water influx. There is still a lack of knowledge of the factors regarding underground water influx and subsidence. However, a more detailed understanding of modeling including the factors for subsidence and water influx can predict the conditions from the output.

Beside the technical and environmental concerns, public perception is very decisive in permitting the process for UCG [109]. Thus, cultivating public acceptance of the technology is another challenge. Recently, the UCG association has been working as an information source of UCG and they are promoting the commercial, social, and environmental benefits of UCG, providing a public and independent information service on aspects of UCG and representing it at the highest levels of governments, regulatory bodies, licensing authorities, energy groups, decision-makers, environmental groups, and the media, maintaining strong links with academia and focusing on education, training, and regulation [110]. As a result, stakeholders are gaining confidence in the new technology and the worldwide professional as well as non-professional communities are getting well educated about UCG technology. They are recognizing UCG as the clean energy source of the future with its outstanding advantages, such as the elimination of coal stock piles and coal transport and much of the disturbance at the surface, low dust and noise levels, the absence of health and safety concerns relating to underground workers, the avoidance of ash handling at power stations, and the elimination of SO_2 and NO_x emissions [111].

Finally, the economics of production and utilisation of the gas plays an important role for commercialization. The primary use of syngas obtained from UCG is to generate electricity; however, in order to achieve an expanded and diversified economy, the syngas are now being used as in many applications, such as the production of chemicals (e.g., hydrogen, ammonia, methanol, ethanol, urea, naptha) or liquid fuels (gasoline, diesel, jet fuel, LPG, *etc.*). In addition, research shows that UCG- IGCC (integrated gasification combined cycle) power plants are cheaper and carbon-efficient [112]. According to the analysis of Brand [113], the cost of syngas production is highly competitive against the available substitute on the market of Central East Europe. The scenario is different for other countries where natural gas is abundant and cheap. The most vital economic factor that delays the commercialization of UCG is the market force—lower natural gas prices. Thus, it is expected that the development of UCG will be in those countries where there is a scarcity of natural gas and an abundance of coal such as in South Africa, China, and Australia, where these developments are taking place with more vigour.

However, at present, the lack of investor confidence and the slower uptake of UCG in the economic climate are due to the absence of the information of a fully operational plant using up-to-date UCG technology [109]. As the worldwide pilot plants of some renowned companies including Linc Energy Ltd., Cougar Energy Ltd. and Carbon Energy Ltd. are working toward the commercialization of UCG and are very close to the commercialization [27], the situation is expected to change very soon in the near future.

4. UCG-CCS (CARBON CAPTURE AND STORAGE)

Currently, sequestration of CO_2 in a subsurface is becoming popular worldwide because of moving towards a low-carbon future. The unminable coal seams are currently being considered as carbon dioxide sinks [111,114]. The accessibility of carbon dioxide to the pores of unworked deep coal seams is limited by the porosity and permeability of coals that tend to decrease with increasing overburden pressure [115]. However, the void created due to gasification and the amount of coal chars left behind after the UCG process triggers the possibility of storing carbon dioxide in that depleted space known as UCG-CCS [116].

Considering the chemistry of gas storage, studies show that gas storage in coal occurs in both the adsorbed phase and the free gas phase. At low pressure, the carbon dioxide storage potential in coal seams is mainly a function of its adsorption capacity; however, at high pressure, it is a strong function of the accessibility to CO_2 with respect to pore space for free gas [115,116,117]. Although the volume of pores in coal is much smaller than in other reservoir rocks, the storage potential of CO_2 in coal significantly differs from other rocks and exceeds its open pore volume by an order of magnitude predominantly due to the surface adsorption of micropores [117]. Saghafi *et al.* [117] reported that coal micropore surface area can reach several hundreds of square meters per gram of solid, which makes large areas available for gas adsorption. However, the development of micropores is a function of coal rank, temperature, and coal composition [117].

There are a number of reports concerning CO_2 storage potential in underground coal seams of different ranks [118]; however, there were inadequate reports of CO_2 storage potential in the coal seams subsequent to gasification. Recently, case studies of the feasibility for UCG-CCS storage have been carried out in the Powder River Basin of Wyoming, US [7,93,113,119], at a selected site in Bulgaria [91,93], and at the Williston Basin, North Dakota, US [108]. Roddy and Younger [111] studied the potential of carbon storage and addressed surface subsidence and aquifer contamination as the main challenges for UCG-CCS application; however, in a later paper, Younger [120] studied hydrogeological and geomechanical aspects of UCG and proposed an action plan in order to avoid subsidence and aquifer contamination in the life cycle of UCG-CCS operations. In order to predict these risks, geomechanical and hydrogeological models are required. The models that dealt with surface subsidence and water hydrology in the UCG process can be found in recent references: Yang *et al.* [97], Sheng *et al.* [98]. From the studies, it appears that careful planning, appropriate site selection, adequate engineering and modeling, and professional management are the keys in order to implement UCG-CCS schemes successfully. It was also understood from the literature that the cavity should be located deeper than 800 m, so that CO_2 can be stored in the supercritical state, allowing significantly higher utilization of the pore space [7,111]. Another interesting feature of UCG-CCS is the cavity temperature during CO_2 injection. As the temperature in the cavity of the UCG process reaches more than 1000 °C, the reactor cool-down process is essential for the preparation of UCG cavities for the subsequent CO_2 storage. The cooling of the cavity can be achieved either naturally very slowly; however, forced cooling with water flushing can significantly decrease the duration of the cool-down process by a factor of more than three hundred [99]. Finally, the implementation of UCG-CCS must be dependent on economic feasibility. Although very few studies are available on the economic assessment for the evaluation of UCG-CCS schemes [98,121], much remains to be done.

5. CONCLUSIONS

A number of significant modeling efforts and the outcomes of these models are presented in this review. The effect of operating conditions, such as temperature, pressure, injected gas composition and flow rate, and steam/O_2 ratio on gas composition or cavity growth rate are revealed by a number of models. For most of the cases, the results are in qualitative agreement among the models. However, it is hard to compare the results quantitatively when the approach of the models and assumptions are different.

The packed bed models are applicable for highly permeable porous media only. Due to the assumption of highly porous media, Darcy's law has been considered for fluid flow. Thus, the possibility of turbulent flow at any location of the gasification channel was ignored. None of the models considered radiation heat transfer and heat loss to the surroundings. However, most of the packed bed models considered a detailed physical and chemical model, based on experimental and theoretical comparisons and correlations. Comprehensive drying and pyrolysis sub-models were also considered in most of the packed bed models. As a result, the packed bed models exhibit good agreement for gas compositions measured from laboratory experiments. In order to calculate the heat recovery and gas composition, the model can be very effective. However, the cavity shape could not be obtained by any of the existing packed bed models due to their 1D framework. Another difficulty is to incorporate thermo-mechanical failure into the models due to their approach. As a result, the extension of these models to field trials is formidable.

The channel models overcome the limitation of the packed bed models in regards to calculating the cavity shape and size. In this method, a permeable channel between the injection and production holes is preferred. Coal is considered to be gasified only at the perimeter of the expanding permeable channel. Natural convection is considered the most important phenomenon for the mixing of the injected blast gas and the gas coming from the channel wall. All models of heat transfer were considered in most of the channel models. Some models successfully incorporated water influx. Fluid flow was not limited to Darcy's law. As a result, in some models, the transport of heat and species inside the cavity was considered to be governed by empirical correlations for turbulent transport phenomena in enclosures. However, most of the channel models neglected drying and pyrolysis, which are considered very important in other approaches of UCG models. Very few models included thermo-mechanical failure. Despite this fact, most of the channel models were validated with the data from field trials. The channel model is found to better calculate sweep efficiency. Most of the existing channel models are in 2D. However, recently, a few 3D models have been developed from which the cavity shape and size are visibly obtained.

The coal slab models describe the process by movement of various defined regions in a coal slab, usually perpendicular to the flow of the injected blast gas. The special feature of the coal slab model is the tracking of drying and combustion front movement. These models can successfully demonstrate the drying and devolatilization behaviour of large coal particles; however, these models are yet to be validated using UCG field trial data. Due to the assumption of semi-infinite coal slab, it can be speculated that this model is only good for thick coal layers. All the models so far were developed by considering that coal slabs are one-dimensional (1D). As a result, information on cavity shape cannot be obtained from these models.

Regardless of the approach of the models, there is not a single model that includes all the important physical and chemical models for the successful prediction of UCG. Although thermo-mechanical failure is a fact for field trials, there are only very few models that consider this phenomenon with unrealistic and simple assumptions. Because of computational time, most of the earlier models are either in a 1D or 2D framework. Even the 3D models developed thus far are very simple and did not consider the details of the physical and chemical models. Considering all these facts about the UCG models, one should follow the approaches of the models depending upon one's objective and available data at hand; for example, one may attempt to follow the approach of the channel model if he/she is interested in modeling the cavity growth, where others may be interested in the packed bed model for gas composition using fairly complicated chemical process and design parameters.

Developing a successful UCG model requires a sound understanding of physical theory as well as knowledge of details to interpret field test and laboratory experimental results in order to successfully predict the field test data and predict the performance of a new site before proceeding with gasification. It is established that each potential UCG site will be unique, so design parameters (*i.e.*, coal seam depth, thickness, inclination, coal properties, moisture content, water influx, *etc.*) are needed to be dealt with first, before operating and performance parameters. That is why a number of experimental laboratory data are needed in order to develop the viable correlation of various properties to use in the model to enhance the accuracy and range of the applicability of the model. A detailed analysis is required for a specific UCG site for seam geology, properties, overburden properties, and hydrology in order to anticipate the possible types of changes in the state of the roof rock so that the approach of modeling can be carefully chosen. The future focus should be the development of a model in a three-dimensional frameworks including all the factors that affect the performance of the UCG.

Commercialization of UCG technology is a vital issue due to some challenges, primarily the environmental concerns and the low energy prices. However, due to recent repeated trial success, progress in handling the environmental issues and the improvement in costs and, finally, the global need for new clean energy sources, UCG is becoming very appealing to the energy sectors. As a result, a number of worldwide renowned companies are working effectively towards commercialization. Considering all these facts, it can be said that UCG is now on the platform close to commercialization. In addition, recent feasibility studies on UCG-CCS schemes added a low-carbon future to UCG.

A number of modeling approaches to gasification and their applicability and relevance have been carefully reviewed. Research in this area is necessary to proceed in order to able to successfully develop and implement the UCG process.

ACKNOWLEDGMENTS

We would like to thank the Canadian Centre for Clean Coal/Carbon and Mineral Processing Technologies (C^{5}MPT), University of Alberta, Canada for the financial support of this work.

AUTHOR CONTRIBUTIONS

M.M. Khan conducted the study and wrote the paper that was started by A.S. Shirazi. Mmbaga also contributed broadly in defining the framework of this paper and revising the manuscript. All other authors contributed by providing their thoughts and suggestion. Finally, the work was carried out under active and direct supervision of Gupta.

CONFLICTS OF INTEREST

The authors declare no conflict of interest.

REFERENCES

1. BP Statistical Review of World Energy—June 2007, Egham, Surrey, UK. BP Distribution Services, p. 45. Available online: http://www.bp.com/content/dam/bp-country/en_ru/documents/publications_PDF_eng/Statistical_review_2007.pdf (accessed on 20 October 2015).
2. Couch, G.R. Underground Coal Gasification; IEA Clean Coal Centre: London, UK, Report No.: Contract No.: CCC/151; 2009; p. 129.
3. Word Energy Council. Survey of Energy Resources, Word Energy Council: London, UK, 2007. Available online: http://ny.whlib.ac.cn/pdf/Survey_of_Energy_Resources_2007.pdf (accessed on 20 October 2015).

4. Tsui, H.; Eu, C. Operating Concept of Circulating Fluidized Bed Gasifier from the Kinetic Point of View.Power Technol. 2003, 132, 176–183.

5. China Labour Bulletin. Deconstructing deadly details from China's coal mine safety statistics CLB. 2006. Available online: http://www.clb.org.hk/en/content/deconstructing-deadly-details-chinas-coal-mine-safety-statistics (accessed on 15 October 2015).

6. US Mine Safety and Health Administration. Statistics—Coal Mining Fatalities by State—Calendar Year. 2015. Available online: http://www.msha.gov/stats/charts/coalbystates.pdf (accessed on 15 October 2015).

7. Shafirovich, E.; Varma, A. Underground Coal Gasification: A Brief Review of Current Status. Ind. Eng. Chem. Res. 2009, 48, 7865–7875.

8. Abdel-Hadi Eed, A.A.; Hsu, T.R. Computer Modeling of Fixed Bed Underground Coal Gasification Using the Permeation Method. J. Energy Resour. Tech. 1987, 109, 11–20.

9. Kariznovi, M.; Nourozieh, H.; Abedi, J.; Chen, Z. Simulation Study and Kinetic Parameter Estimation of Underground Coal Gasification in Alberta Reservoirs. Chem. Eng. Res. Des. 2013, 91, 464–476.

10. Bhutto, A.W.; Bazmi, A.A.; Zahedi, G. Underground Coal Gasification: From Fundamentals to Applications.Prog. Energy Combust. Sci. 2013, 39, 189–214.

11. Burton, E.; Friedmann, J.; Upadhye, R. Best Practices in Underground Coal Gasification; Lawrence Livermore National Laboratory: California, USA, 2006. Available online: http://www.purdue.edu/discoverypark/energy./pdfs/cctr/BestPracticesinUCG-draft.pdf (accessed on 13 August 2015).

12. Nourozieh, H.; Kariznovi, M.; Chen, Z.; Abedi, J. Simulation Study of Underground Coal Gasification in Alberta Reservoirs: Geological Structure and Process Modeling. Energy Fuels 2010, 24, 3540–3550.

13. Anthony, D.B.; Howard, J.B. Coal Devolatilization and Hydrogastification. AIChE J. 1976, 22, 625–656.

14. Yang, L. Numerical Study on the Underground Coal Gasification for Inclined Seams. AIChE J. 2005, 51, 3059–3071.

15. Perkins, G. Mathematical Modelling of Underground Coal Gasification. Ph.D. Thesis, The University of New South Wales, Sydney, Australia, 2005.

16. Klimenko, A.Y. Early Ideas in Underground Coal Gasification and Their Evolution. Energies 2009, 2, 456–476.

17. Kumar, R.N.H.; Udayakumar, D.L.; Stojcevski, A.; Maung Than Oo, A.M. Underground Coal Gasification: An alternate, Economical, and Viable Solution for future Sustainability. Int. J. Eng. Sci. Invent. 2014, 3, 57–68.

18. Su, F.; Nakanowataru, T.; Itakura, K.-I.; Ohga, K.; Deguchi, G. Evaluation of Structural Changes in the Coal Specimen Heating Process and UCG Model Experiments for Developing Efficient UCG Systems. Energies2013, 6, 2386–2406.

19. Gregg, D.W.; Edgar, T.G. Underground Coal Gasification. AIChE J. 1978, 24, 753–781.

20. Skafa, P.V. Underground Gasification of Coal; Gosudarstvennoe Hauchno-Tekhnicheskoe Izdatel'stvo Literatury Po Gornomu Delu, Moscow, 1960; p. 324; English transl.: Report No. UCRL-Trans 10880; Lawrence Livermore National Laboratory: Livermore, CA, USA, 1975.

21. Gregg, D.W.; Hill, R.W.; Olness, D.U. An Overview of the Soviet Effort in Underground Coal Gasification; Report No. UCRL-52004; Lawrence Livermore National Laboratory: Berkeley, CA, USA, 1976.

22. Derbin, Y.; Walker, J.; Wanatowski, D.; Marshall, A. Soviet Experience of Underground Coal Gasification Focusing on Surface Subsidence. Appl. Phys. Eng. 2015, 16, 839–850.

23. Thorsness, C.B.; Britten, J.A. Underground Coal Gasification Project: Final Report; Report No. UCID-21853; Lawrence Livermore National Laboratory: Livermore, CA, USA, 1989.

24. Blinderman, M.S.; Saulov, D.N.; Klimenko, A.Y. Forward and Reverse Combustion Linking in Underground Coal Gasification. Energy 2008, 33, 446–454.

25. Cui, Y.; Liang, J.; Wang, Z.; Zhang, X.; Fan, C.; Liang, D.; Wang, X. Forward and Reverse Combustion Gasification of Coal with Production of High-Quality Syngas in a Simulated Pilot System or In SituGasification. Appl. Energy 2014, 131, 9–19.

26. Bedi, P.; Binshtok, A.; Blinderman, A.; Blinderman, M.; Elliott, G. Application of the Exergy UCG Technology in CTX Projects. In Proceedings of the 2nd Underground Coal Gasification Network Workshop, Banff, AB, Canada, 22–23 August 2012.

27. Shirazi, A.A. CFD Simulation of Underground Coal Gasification. Master's Thesis, University of Alberta, Alberta, Canada, 2012.

28. Davis, B.; Mallett, C. Bloodwood Ck UCG Developments. In Proceedings of the 2nd Workshop-Underground Coal Gasification Network, Banff, AB, Canada, 22–23 August 2012.

29. Beath, A.; Craig, S.; Littleboy, A.; Mark, R.; Mallett, C. Underground Coal Gasification: Evaluating Environmental Barriers. CSIRO Exploration and Mining Report P2004/5, Kenmore, Queensland, Australia. 2004.

30. Shanon, M.J.; Thorsness, C.B.; Hill, R.W. Early Cavity Growth during forward Burn; Report No. UCRL-84584; Lawrence Livermore National Laboratory: Livermore, CA, USA, 1980.

31. Hill, R.W. Burn Cavity Growth during the Hoe Creek No. 3 Underground Coal Gasification Experiment; Report No. UCRL-85173-R-1, U.S. DOE; Lawrence Livermore National Laboratory: Livermore, CA, USA, 1981.

32. Yeary, D.L.; Riggs, J.B. Study of Small-scale Cavity Growth Mechanisms for Underground Coal Gasification.In Situ 1987, 11, 305–327.

33. Poon, S.S. The Combustion Rates of Texas Lignite Cores. Masters Thesis, University of Texas, Austin, TX, USA, 1985.

34. Daggupati, S.; Mandapati, R.N.; Mahajani, S.M.; Ganesh, A.; Mathur, D.K.; Sharma, R.K.; Aghalayam, P. Laboratory Studies on Combustion Cavity Growth in Lignite Coal Blocks in The Context of Underground Coal Gasification. Energy 2010, 35, 2374–2386.

35. Yang, L. Study on the Method of Two-Phase Underground Coal Gasification with Unfixed Pumping Points.Energy Sources 2003, 25, 917–930.

36. Prabu, V.; Jaynati, S. Simulation of Cavity Formation in Underground Coal Gasification using Borehole Combustion. Energy 2011, 36, 5854–5864.

37. Stańczyk, K.; Howaniec, N.; Smoliński, A.; Świądrowski, J.; Kapusta, K.; Wiatowski, M.; Grabowski, J.; Rogut, J. Gasification of Lignite and Hard Coal with Air and Oxygen Enriched Air in A Pilot scale Ex Situ Reactor for Underground Gasification. Fuel 2011, 90, 1953–1962.

38. Thorsness, C.B.; Hill, R.W. Coal Block Gasification: Laboratory Results and Field Plans; Report No. UCRL-85906; Lawrence Livermore National Laboratory: Livermore, CA, USA, 1981.

39. Upadhye, R.; Burton, E.; Friedmann, J. Science and Technology Gaps in Underground Coal Gasification; Report No. UCRL-TR-222523; Lawrence Livermore National Laboratory: Livermore, CA, USA, 2006.

40. Gunn, R.D.; Krantz, W.B. Underground Coal Gasification: Development of Theory, Laboratory Experimentation, Interpretation, & Correlation with the Hanna Field Tests: Final report; U.S. Department of Energy: Morgantown, WV, USA, 1987.

41. Higgins, G.H. A New Concept for in-Situ Coal Gasification; Report No. UCRL-51217; Lawrence Livermore National Laboratory: Livermore, CA, USA, 1976.

42. Thorsness, C.B.; Rosza, R.B. In-Situ Coal Gasification Program: Model Calculations and Laboratory Experiments; Report No. UCRL-78302; Lawrence Livermore National Laboratory: Livermore, CA, USA, 1976.

43. Uppal, A.A.; Bhatti, A.I.; Aamir, E.; Samar, R.; Khan, S.A. Control Oriented Modeling and Optimization of One Dimensional Packed Bed Model of Underground Coal Gasification. J. Process Control 2014, 24, 269–277.

44. Dinsmoor, B.; Galland, J.M.; Edgar, T.F. The Modeling of Cavity Formation during Underground Coal Gasification. J. Petroleum Technol. 1978, 30, 695–704.

45. Gunn, R.D.; Whitman, D.L. An In-Situ Coal Gasification Model (Forward Mode) for Feasibility Studies and Design; Report No. LERC/RI-76/2; Laramie Energy Research Center: Laramie, WY, USA, 1976.

46. Khadse, A.N.; Qayyumi, M.; Mahajani, S.; Aghalayam, P. Model for the Underground Coal Gasification (UCG) Channel Reactor Model for the Underground Coal Gasification (UCG) Channel. Int. J. Chem. React. Eng. 2006, 4, 1–25.

47. Winslow, A.M. Numerical Model of Coal Gasification in A Packed Bed; Report No. UCRL-77627; Lawrence Livermore National Laboratory (LLNL): Berkeley, CA, USA, 1976.

48. Thorsness, C.B.; Grens, E.A.; Sherwood, A.A. One Dimensional Model for in Situ Coal Gasification; Report No. UCRL-52523; Lawrence Livermore National Laboratory: Livermore, CA, USA, 1978.

49. Campbell, J.H. Pyrolysis of Subbituminous Coal in Relation to In-Situ Coal Gasification. Fuel 1978, 57, 217–224.

50. Shirazi, A.S.; Karimipour, S.; Gupta, R. Numerical Simulation and Evaluation of Cavity Growth in In SituCoal Gasification. Ind. Eng. Chem. Res. 2013, 52, 11712–11722.

51. Thorsness, C.B.; Kang, S.W. Further Development of a General-Purpose Packed Bed Mode for Analysis of UCG Process; Report No. UCRL-92489; Lawrence Livermore National Laboratory: Berkeley, CA, USA, 1985.

52. Harker, J.H.; Backhurst, J.R. Fuel and Energy (Energy Science and Engineering); Academic Press: London, UK, 1981.

53. Gadsby, J.; Hinshelwood, C.N.; Sykes, K.W. The Kinetics of the Reactions of the Steam-Carbon System. Proc. R. Soc. (Lond.) Series A (Math. Phys. Sci.) 1946, 187, 129–151.

54. Lewis, W.K.; Gilliland, E.R.; Paxton, R.R. Low-Temperature Oxidation of Carbon. Ind. Eng. Chem. 1954, 46, 1327–1331.

55. Batenburg, W.D.; Biezen, N.J.; Bruining, J. A New Channel Model for Underground Coal Gasification of Thin, Deep Coal Seams. In Situ 1994, 18, 419–451.

56. Kuyper, R.A.; van der Meer, T.H.; Bruining, J. Simulation of Underground Gasification of Thin Coal Seams.In Situ 1996, 20, 311–346.

57. Eddy, T.L.; Schwartz, S.H. A Side Wall Burn Model for Cavity Growth in Underground Coal Gasification. J. Energy Resour. Tech. 1983, 105, 145–156.

58. Luo, Y.; Coertzen, M.; Dumble, S. Comparison of UCG Cavity Growth with CFD model predictions. In Proceedings of the 7th International Conference on CFD in the Minerals and Process Industries CSIRO, Melbourne, Australia, 9–11 December 2009.

59. Pirlot, P.; Pirard, J.P.; Coeme, A.; Mostade, M.A. Coupling of Chemical Processes and Flow in View of the Cavity Growth Simulation of an Underground Coal Gasifier at Great Depth. In Situ 1998, 22, 141–156.

60. Kuyper, R.A.; van der Meer, T.H.; Hoogendoorn, C.J. Turbulent Natural Convection Flow due to Combined Buoyancy Forces during Underground Coal Gasification of Thin Seams. Chem. Eng. Sci. 1994, 49, 851–861.

61. Perkins, G.; Sahajwalla, V. Modelling of Heat and Mass Transport Phenomena and Chemical Reaction in Underground Coal Gasification. Chem. Eng. Res. Des. 2007, 85, 329–343.

62. Schwartz, S.Y.; Eddy, T.L.; Mehta, K.H.; Lutz, S.A.; Binaie-Kondolojy, M.B. Cavity growth mechanisms in UCG with side wall burn gasification. In Proceedings of the SPE Annual Fall Technical Conference and Exhibition, Houston, TX, USA, 1–3 October 1978.

63. Daney, D.E. Turbulent Natural Convection of Liquid Deuterium Hydrogen, and Nitrogen within Enclosed Vessels. Int. J. Heat Mass Transf. 1970, 19, 431–441.

64. Bailey, A.C.; Bartlett, F.; Boswinkel, H.H.; Bruining, J.; Chandelle, V.; Del Amor, G.; Garstang, J.H.; Hewel-Bundermann, H.; Hewing, G.;

Krabiell, K.; et al. The Future Development of Underground Coal Gasification in Europe; A Comprehensive Report to CEC; Commission of European Communities: Brussels, Belgium, 1989.

65. Wilks, I.H.C. The Cavity Produced by Gasifying Thin Deep Seams. In Proceedings of the 9th Underground Coal Gasification Symposium US, Washington, DC, USA, 7 August 1983; pp. 314–322.

66. Cena, R.J.; Britten, J.A.; Thorsness, C.B. Excavation of the Partial Seam CRIP Underground Coal Gasification Test Site; Report No. 97245; Lawrence Livermore National Laboratory: Livermore, CA, USA, 1987.

67. Bruining, J.; Herkstroter, N.M.; Lankhorst, A.M.; Visser, C.P.; Ronde, H. Free Convection Enhanced Mass Transfer during Underground Tunnel Gasification of Coal. In Proceedings of the Thirteenth Annual Underground Coal Gasification Symposium; U.S. Department of Energy: Washington, DC, USA, 1987; pp. 225–233.

68. Hill, R.W.; Thorsness, C.B. Summary Report on Large Block Experiments in Underground Coal Gasification, Tono Basin, Washington: Volume 1. Experimental Description and Data Analysis; Report No. UCRL-53305; Lawrence Livermore National Laboratory (LLNL) Report: Berkeley, CA, USA, 1982.

69. Oliver, R.L.; Mason, G.M.; Spackman, L.K. Field and Laboratory Results from the TONO I (CRIP) UCG Cavity Excavation Project, WIDCO Mine Site, Centralia, Washington. Fuel Sci. Tech. 1989, 7, 1059–1120.

70. Launder, B.E.; Sharma, B.I. Application of the Energy Dissipation Model of Turbulence to the Calculation of the Flow near a Spinning Disc. Lett. Heat Mass Transfer. 1974, 1, 131–138.

71. Biezen, E.N.J. Modeling Underground Coal Gasification. Ph.D. Thesis, Delft University of Technology, Delft, The Netherlands, 1996.

72. Saulov, D.N.; Plumb, O.A.; Klimenko, A.Y. Flame Propagation in a Gasification Channel. Energy 2010, 35, 1264–1273.

73. Chernyshev, A.B. Izbrannye Trudy; Izdatel'stvo: Moscow, USSR, 1956.

74. Tsang, T.H.T. Modeling of Heat and Mass Transfer during Coal Block Gasification. Ph.D. Thesis, University of Texas at Austin, Austin, TX, USA, 1980.

75. Westmoreland, P.R.; Forrester, R.C., III. Pyrolysis of Large Coal Blocks: Implications of Heat and Mass Transport Effects for In-Situ Gasification. ACS Div. Fuel Chem. 1977, 22, 93.

76. Perkins, G.; Sahajwalla, V. A Numerical Study of the Effects of Operating Conditions and Coal Properties on Cavity Growth in Underground Coal Gasification. Energy 2006, 20, 596–608.

77. Massaquoi, J.G.M.; Riggs, J.B. Mathematical Modeling of Combustion and Gasification of a Wet Coal Slab—I: Model Development and Verification. Chem. Eng. Sci. 1983, 38, 1747–1756.

78. Massaquoi, J.G.M.; Riggs, J.B. Mathematical Modeling of Combustion and Gasification of a Wet Coal Slab—II: Mode of Combustion, Steady State Multiplicities and Extinction. Chem. Eng. Sci. 1983, 38, 1757–1766.

79. Park, K.Y.; Edgar, T.F. Modeling of Early Cavity Growth for Underground Coal Gasification. Ind. Eng. Chem. Res. 1987, 6, 237–246.

80. Perkins, G.; Sahajwalla, V. A Mathematical Model for the Chemical Reaction of a Semi-infinite Block of Coal in Underground Coal Gasification. Energy Fuels 2005, 19, 1679–1692.

81. Forrester, R.C. Two-Dimensional Studies of Coal Pyrolyis Preliminary Results. In Proceedings of the 5th Annual Underground Coal Conversion Symposium; U.S. Department of Energy: Washington, DC, USA, 1979.

82. Wellborn, T.A. Linear Burning Rates of Texas Lignite. Master's Thesis, University of Texas, Austin, TX, USA, 1981.

83. Tsang, T.H.; Edgar, T.F. Modeling of Drying and Pyrolysis during Underground Coal Gasification—I. In Situ1983, 7, 237–264.

84. Hsia, S.J.; Wellborn, T.A.; Edgar, T.F. Measurement of Combustion Rates of Texas Lignite with Application to Underground Coal Gasification. In Proceedings of the Fourth UCC Symposium, CO, USA, 17–20 July 1978.

85. Poon, S.S.M.; Edgar, T.F.; Park, K.Y. The Combustion Rates of Texas Lignite Cores. In Situ 1986, 10, 109–143.

86. Chang, H.L.; Himmelblau, D.M.; Edgar, T.F. A Sweep Efficiency Model for Underground Coal Gasification.In Situ 1985, 9, 185–221.

87. Debelle, B.; Malmendier, M.; Mostade, M.; Pirard, J.P. Modelling of Flow at Thulin Underground Coal Gasification Experiments. Fuel 1992, 71, 95–106.

88. Thorsness, C.B. Steam Tracer Experiment at the Hoe Creek No.3 Underground Coal Gasification Field Test; Report No. UCRL-53082; Lawrence Livermore National Laboratory: Livermore, CA, USA, 1980.

89. Pirard, J.P.; Brasseur, A.; Coeme, A.; Mostade, M.; Pirlot, P. Results of the Tracer Tests during the El Tremedal Underground Coal Gasification at Great Depth. Fuel 2000, 79, 471–478.

90. Daggupati, S.; Mandapati, R.N.; Mahajani, S.M.; Ganesh, A.; Pal, A.K.; Sharma, R.K.; Aghalayam, P. Compartment Modeling for Flow Characterization of Underground Coal Gasification Cavity. Ind. Eng. Chem. Res. 2011, 50, 277–290.

91. Batenburg, W.D.; Bruining, J. The Efficiency of Filtration Gasification. In Situ 1993, 17, 413–437.

92. Biezen, E.N.; Bruining, J.; Molenaar, J. An Integrated 3D Model for Underground Coal Gasification. In Proceedings of the Annual Technical Conference of the Society of Petroleum Engineers, Dallas, TX, USA, 22–25 October 1995.

93. Biezen, E.N.; Bruining, J.; Molenaar, J. An Integrated Model for Underground Coal Gasification. In Proceedings of the Annual Technical Conference of the Society of Petroleum Engineers, New Orleans, LA, USA, 25–28 September 1994.

94. Seifi, M.; Chen, Z.; Abedi, J. Numerical Simulation of Underground Coal Gasification using the CRIP method. Can. J. Chem. Eng. 2011, 89, 1528–1535.

95. Nitao, J.J.; Camp, D.W.; Buscheck, T.A.; White, J.A.; Burton, G.C.; Wagoner, J.L.; Chen, M. Progress on a New Integrated 3-D UCG Simulator and its Initial Application. LLNL-CONF-500551. In Proceedings of the International Pittsburgh Coal Conference, Pittsburgh, PA, USA, 12–15 September 2011.

96. Srikantiah, S.; Samdani1, G.; Mahajani1, S.; Ganesh, A.; Aghalayam, A. 2-D Modeling of Underground Coal Gasification (UCG). In Proceedings of the Comsol Conference, Bangalore, India, 17 October 2013.

97. Yang, D.; Sarhosis, V.; Sheng, Y. Thermal-mechanical Modelling around the Cavities of Underground Coal Gasification. J. Energy Inst. 2014, 87, 321–327.

98. Sheng, Y.; Benderev, A.; Bukolska, D.; Eshiet, K.I.; da Gama, C.D.; Gorka, T.; Green, M.; Hristov, N.; Katsimpardi, I.; Kempka, T.; et al. Interdisciplinary Studies on the Technical and Economic Feasibility of Deep Underground Coal Gasification with CO2 Storage in Bulgaria. Mitig. Adapt. Strateg. Glob. Chang.2015, 1–33.

99. Sarhosis, V.; Yang, D.; Sheng, Y.; Kempka, T. Thermal Analysis of Underground Coal Gasification Rector Cool Down for Subsequent CO2 Storage. Energy Procedia 2013, 40, 428–436.

100. Perkins, G.; Sahajwalla, V. Steady-State Model for Estimating Gas Production from Underground Coal Gasification. Energy Fuels 2008, 22, 3902–3914.

101. Lyczkowski, R.W.; Chao, Y.T. Comparison of Stefan Model with Two-Phase Model of Coal Drying. Int. J. Heat Mass Transfer 1984, 27, 1157–1169.

102. Lyczkowski, R.W. Mechanistic Theory for Drying of Porous Media; Report No.: UCRL-52456; Lawrence Livermore National Laboratory: Livermore, CA, USA, 1978.

103. Suuberg, E.M.; Peters, W.A.; Howard, J.B. Product Composition and Kinetics of Lignite Pyrolysis. I.E.C. Process Des. Dev. 1978, 17, 37–46.

104. Boroson, M.L.; Howard, J. Product Yields and Kinetics from Vapor Phase Cracking of Wood Pyrolysis Tars.AIChE J. 1989, 35, 120–128.

105. Juntgen, H.; van Heek, K.H. Gas Release from Coal as a Function of the Rate of Heating. Fuel 1968, 47, 103–117.

106. Van Krevelen, D.W.; van Heerden, C.; Huntjens, F.J. Physiochemical Aspects of the Pyrolysis of Coal and Related Organic Compounds. Fuel 1951, 30, 253–259.

107. Liu, H.; Yao, H.; Yao, K.; Chen, F.; Luo, G. Characteristics of "Three zones" during Underground Coal Gasification. Adv. Mater. Res. 2012, 524–527, 56–62.

108. Gunn, R.D. Problem Solved and Problem not Solved in UCG. Laramie Energy Research Center, Laramie, Wyoming. Preprints of Papers—American Chemical Society. Div. Fuel Chem. 1977, 22, 64–75.

109. Green, M. Underground coal gasification: On the road to commercialisation. Mod. Power Syst. 2014, 34, 22–24.

110. Lauder, J. Global Development of UCG. UCG Association. In Proceedings of the 2nd Underground Coal Gasification Network Workshop, Banff, AB, Canada, 22–23 August 2012.

111. Roddy, D.J.; Paul, L.; Younger, P.L. Underground Coal Gasification with CCS: A Pathway to Decarbonising Industry. Energy Environ. Sci. 2010, 3, 400–407.

112. Blinderman, M.S.; Jones, R.M. The Chinchilla IGCC Project to Date: Underground coal gasification and Environment. In Proceedings of the Gasification Technologies Conference, San Francisco, CA, USA, 27–30 October 2002.

113. Brand, J. Progress Feedback on Wildhorse's Hungarian UCG Project. Wildhorse Energy. In Proceedings of the 2nd Underground Coal

Gasification Network Workshop, Banff, Alberta, Canada, 22–23 August 2012.

114. Pei, P.; Zeng, Z.; He, J. Feasibility Study of Underground Coal Gasification Combined with CO2 Capture and Sequestration in Williston Basin, North Dakota. In Proceedings of the 44th US Rock Mechanics Symposium and 5th U.S.—Canada Rock Mechanics Symposium, Salt Lake City, UT, USA, 27–30 June 2010.

115. Kempka, T.; Fernandez-Steeger, T.; Li, D.-Y.; Schulten, M.; Schluter, R.; Krooss, B.M. Carbon Dioxide Sorption Capacities of Coal Gasification Residues. Environ. Sci. Technol. 2011, 45, 1719–1723.

116. Ramasamy, S.; Sripada, P.P.; Khan, M.M.; Tian, S.; Trivedi, J.; Gupta, R. Adsorption Behavior of CO2 in Coal and Coal Char. Energy Fuels 2014, 28, 5241–5251.

117. Saghafi, A.; Faiz, M.; Roberts, D. CO2 Storage and Gas Diffusivity Properties of Coals from Sydney Basin, Australia. Int. J. Coal Geol. 2007, 70, 240–254.

118. Goodman, A.L.; Campus, L.M.; Schroeder, K.T. Direct Evidence of Carbon Dioxide Sorption on Argonne Premium Coals Using Attenuated Total Reflectance-Fourier Transform Infrared Spectroscopy. Energy Fuels2005, 19, 471–476.

119. Zamzow, K.L. Underground Coal Gasification: History, Environmental Issues, and the Proposed Project at Beluga, Alaska. Available online: http://www.groundtruthtrekking.org/Documents/UCG/UCG-KZamzow-2010.pdf (accessed on 15 October 2015).

120. Younger, P. Hydrogeological and Geomechanical Aspects of Underground Coal Gasification and its Direct Coupling to Carbon Capture and Storage. Mine Water Environ. 2011, 30, 127–140.

121. Nakaten, N.; Schlüter, R.; Azzam, R.; Kempka, T. Development of a Techno-Economic Model for Dynamic Calculation of COE, Energy Demand and CO2 Emissions of an Integrated UCG–CCS Process. Energy 2014,66, 779–790.

8

IMPACT OF UNDERGROUND MINING ON SHAFT LINING AND AQUIFER IN EASTERN CHINA

Qing Yu, Hideki Shimada, Takashi Sasaoka, Kikuo Matsui

Department of Earth Resources Engineering, Faculty of Engineering, Kyushu University, Fukuoka, Japan

ABSTRACT

Serious shaft lining ruptures have often occurred in the eastern part of China since 1987 due to the complicated geological conditions. This paper tries to find out the relationship between mechanisms of shaft lining rupture and the underground mining process. The analysis is based on the existence typical engineering and geological conditions in eastern China; the impact of underground mining on the shaft lining and aquifer layer is analyzed by using numerical method. The impact factors such as different depths, thicknesses, mining widths of coal seam and different distances to the shaft are used in the analysis. The mining area under the aquifer which near the shaft lining has a significant impact on the shaft lining due to mining process, and increases the risk of occurrence of shaft lining rupture.

Keywords: shaft lining; impact; underground mining; aquifer; numerical simulation

1. INTRODUCTION

Serious shaft lining ruptures have often occurred in the eastern part of China and this problem has been one of the major subjects in Chinese coal mine industry for past twenty years. All these sites that shaft lining rupture occurred are in similar geological conditions. All these shafts pass through deep alluvium of Quaternary strata which the composition of the bottom aquifer is complex [1].

According to the statistical data of the shaft lining rupture in alluvium, the elapsed time after drivaging and location of rupture has its own rules. Generally, the type of shaft lining ruptures is divided into three phases [2]: 1) Rupture during the freezing shaft sinking: horizontal cracks caused by temperature variation is initiated and propagated; 2) Frozen wall melting: horizontal and inclined cracks caused by temperature variation are initiated and propagated; 3) Surface subsidence after several years: horizontal and inclined cracks caused by surface subsidence are propagated, especially, this phenomena is occurred near the boundary between aquifer and bedrock.

The shaft lining rupture happening after several years running is the most common situation in the eastern part of China, such as Datun, Xuzhou, Huaibei, Yanzhou, Yingxia, Hebi, Dongrong, etc., since 1987 [3-5]. When shaft lining rupture occurs, the inner shaft lining delaminates and spalls, longitudinal steel bows inward, transverse cracks form and intersect in the horizontal direction along circle, seepage occurs or even sand gushes, and mostly seriously, concrete blocks fall out and break the equipment in shaft. In addition, the shaft bends up; cage guides, drainpipes and pressure ventilation pipes are in longitudinal bending, and in serious cases the cage is stuck due to torsional deformation.

According to several surveys on coal mines, before the shaft lining rupture happening, the groundwater leakage has been observed in the mining area from roof, the groundwater level decline is happening at the same time in the aquifer [6-9]. The current researches are mostly focused on the causes, mechanisms and solutions for these kinds of geotechnical issues and they have been the major topics in recent 20 years. In this research, the model of strata layers is built and used for numerical simulation, and the impact of coal mining in the deeper coal seam before aquifer drawdown on shaft lining is examined.

2. ANALYSIS MODEL

FLAC^{3D} is a three-dimensional explicit finite-difference program for engineering mechanics computation and simulating the behavior of three-dimensional structures built of soil, rock or other materials that undergo plastic flow when their yield limits are reached.

The basic model used in the analysis is shown in **Figures 1** and **2**. This model represents the typical engineering and geological conditions in eastern China especially Baodian coal mine which happened the shaft lining rupture in 1995. The net diameter of the main shaft in

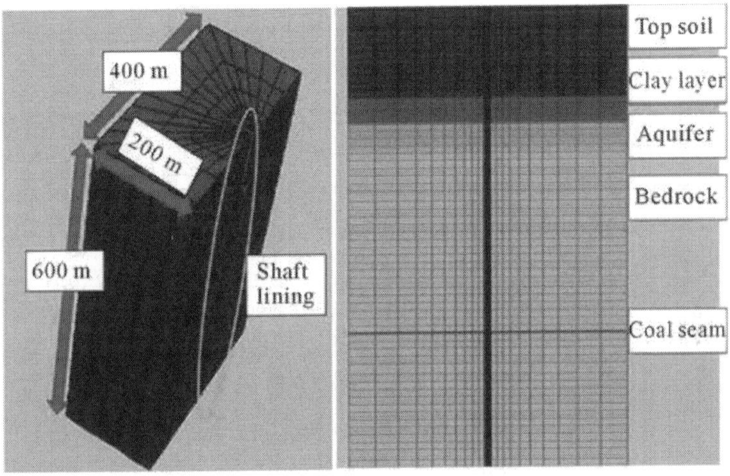

Figure 1. Analysis model.

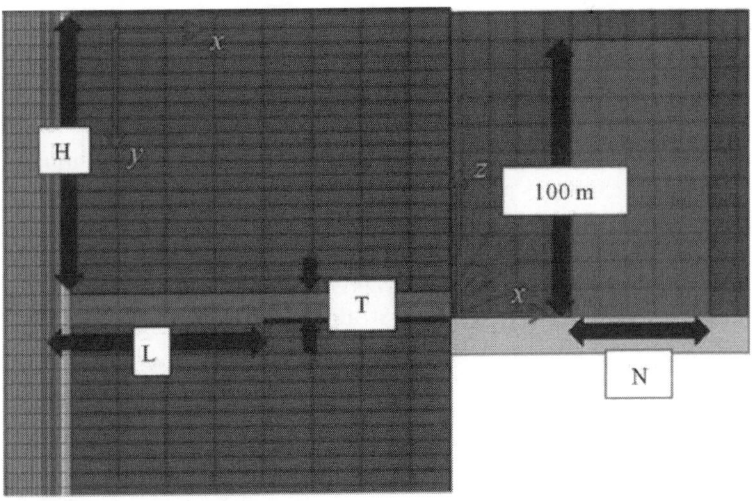

Figure 2. Mining area model.

this mine is 6.5 m and the thickness of the shaft lining is 1.0 m constructed by C28 concrete. The thickness of alluvium is 148.69 m and it contains 3 aquifers. The bottom aquifer is important to the mine and the water level is affected by mining, so in this model, only the bottom aquifer is left for simulation.

According to the engineering geological data and shaft construction data of coal mine, the required mechanical parameters for numerical simulation are shown in **Table 1**.

As the process of soil deformation and shaft lining stress change is very complex, the following assumptions are made:

- Shaft lining, soil and loss areas are treated as the symmetrical distribution, belong to the space axialsymmetry.
- Concrete shaft lining, the surrounding soil and aquifer are homogeneous, isotropic. No interface between the surrounding soil and shaft lining, that means no sliding in the interface.

Then the boundary conditions are set as follows: a free top surface, two horizontal fixed slides and a vertical fixed the lower surface.

As the coal seam is under the deep bottom aquifer, the different depth, thickness, mining width of coal seam and different distance to the shaft were simulated as the impact factors of the shaft lining and the aquifer (as shown in **Table 2**).

3. ANALYSIS RESULTS AND DISCUSSION

3.1. Failure Initiation Due to Mining Process

The shear strain change is chosen to reveal the impact of the mining process, as the rock compression failure mainly belongs to the shear failure [10]. **Figure 3** shows the shear strain distribution around the mining area and the aquifer during mining. The high strain area is in yellow colour and the low strain area is in blue. From this figure, the both ends of the mining area and the aquifer layer have the high strain. In addition, the shear strain in the rock layer between the mining area and the aquifer exceeds 2.0×10^{-3}. And except the part shown in blue, there is a risk of failure occurs, especially the area between mining area and aquifer. Therefore, the leakage occurs at the impermeability rock layer can be concluded by the fissure developing in the rock which the failure happened during the mining process. According to the above analysis, during the underground mining in 250 m depth, the bedrock under aquifer begins to get loose and then some fissure develops, as a result the aquifer drainage occurs.

Figures 4 and **5** show the distribution of shear strain in different mining depth. The deeper of the mining area, the distance to the aquifer is greater, then the impact on the aquifer is gradually reduced. The impact on the aquifer becomes disappear while mining in 400 m depth, and the shear strain near the mining area tends to become higher. For this reason, if the mining depth is deeper than 400 m, it becomes much more important to consider the safety of the surrounding area compare with the impact on the aquifer.

Table 1. Mechanical parameters of soil and shaft lining.

Material	Top soil	Clay	Aquifer	Bedrock	Shaft lining	Coal seam
Thickness (m)	120	30	30	450	1	-
Dry density (g/cm^3)	1.65	1.65	1.7	2.7	3.0	2.7
Bulk modulus (Mpa)	50	50	61.25	1.1×10^4	2×10^4	1.1×10^4
Shear modulus (Mpa)	23	23	28	8333	12600	8333
Porosity	0.18	0.05	0.3	-	-	-

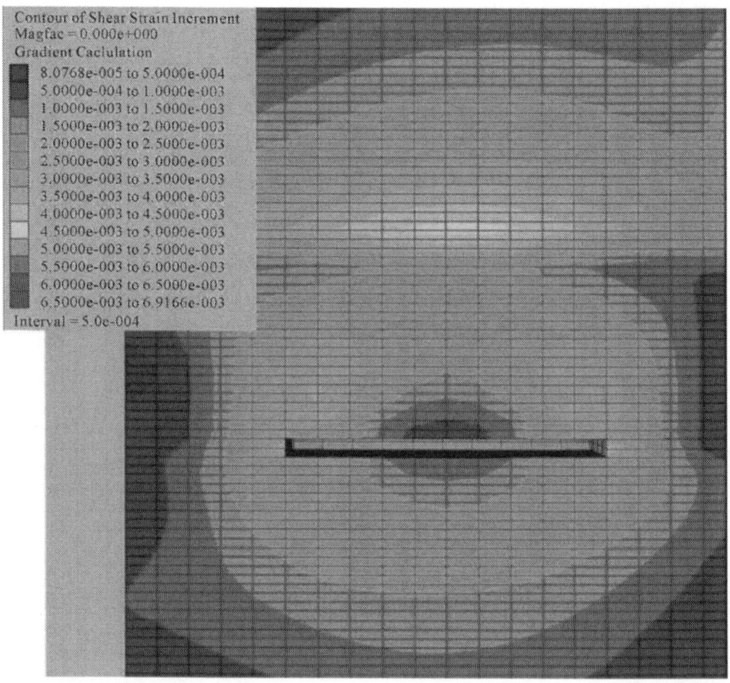

Figure 3. Shear strain distribution during mining (50 m to the shaft and 250 m mining depth, 6 m thickness coal seam).

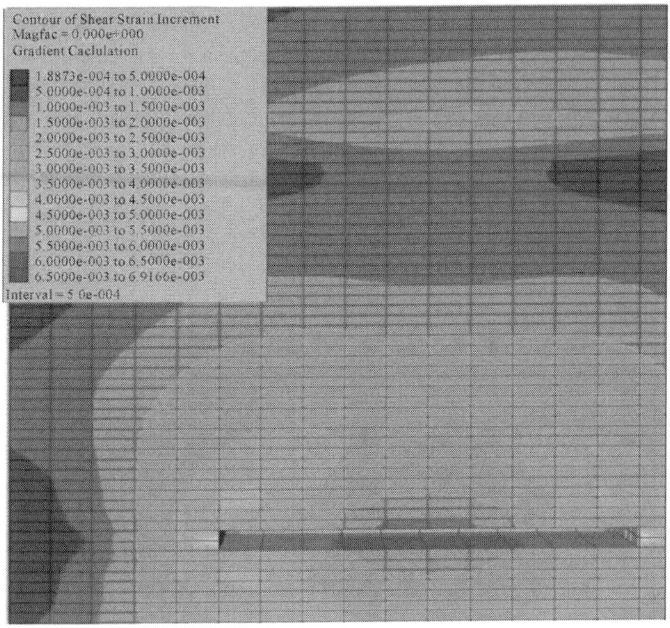

Figure 4. Shear strain distribution during mining (50 m to the shaft and 300 m mining depth, 6 m thickness coal seam).

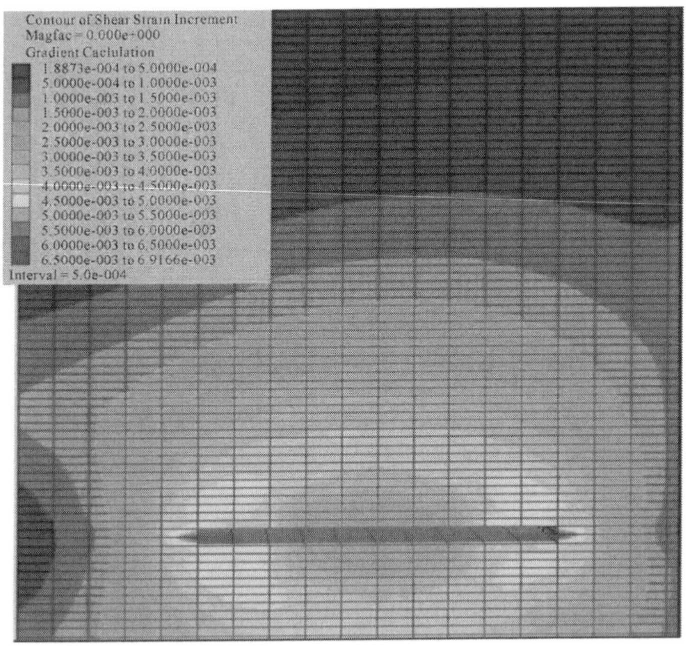

Figure 5. Shear strain distribution during mining (50 m to the shaft and 400 m mining depth, 6 m thickness coal seam).

Table 2. Factors for analysis.

Depth of coal seam (H)	250 m × 300 m × 400 m
Thickness of coal seam (T)	3 m × 6 m × 9 m
Mining width (N)	50 m × 100 m
Distance to shaft (L)	50 m × 70 m × 90 m

3.2. Impact Due to Mining Distance to Shaft

The stress in shaft lining is the most concerned issue, so the impact to the shaft lining is revealed by stress change. **Figure 6** shows the impact on the surrounding areas and the shaft lining by different distance of mining area to the shaft. The lines in the figure are the values of maximum principal stress in the surrounding area, aquifer and shaft lining. As the distance between mining area and shaft becomes greater, the maximum principal stress in shaft lining becomes smaller, the mining impact on the shaft lining becomes smaller and that can reduce the concentration of the maximum stress.

3.3. Impact Due to Mining Depth

The mining depth also has the impact to the stress in the shaft lining, though **Figure 7** shows the maximum principal stress change due to different mining depths. As the mining area is deeper and the overburden increases, the deformation of surrounding area is greater and the stress concentration in mining area and shaft lining tends to rise as the depth increases. Also the maximum principal stress in the mining area is higher than in the shaft lining. From the above analysis, the risk of shaft lining rupture is smaller than the risk of mining area collapse. So the coal pillar and artificial support need to be remained in the mining area considering the collapse risks.

3.4. Impact Due to Mining Width

The change of mining width leads to the deformation of the ground layers; the displacement of the ground layers is used to show the impact. **Figures 8** and **9** show the influence of 50 m and 100 m mining width. As the mining width increases, the deformation of surrounding mining area increases which it also gives an impact on the deformation of the shaft lining. As shown in **Figure 10**, when the mining width is 50 m, the maximum principal stress concentration occurs in the surrounding area, and when the mining width is 100 m, the maximum stress occurs in the shaft lining.

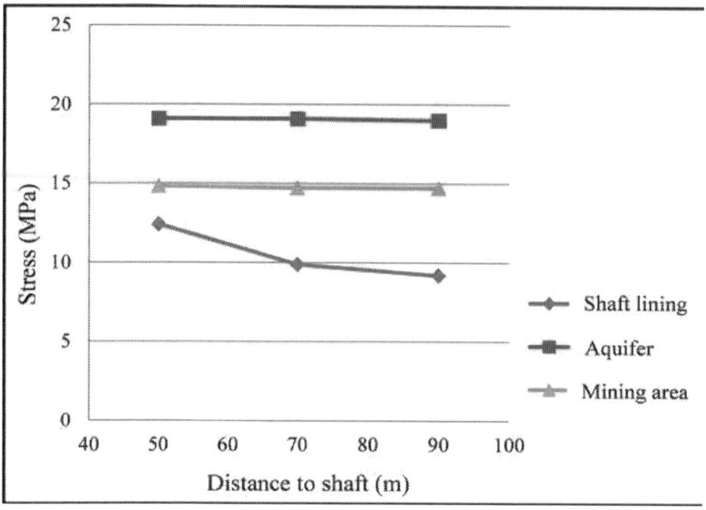

Figure 6. Impact due to distance to shaft (250 m mining depth).

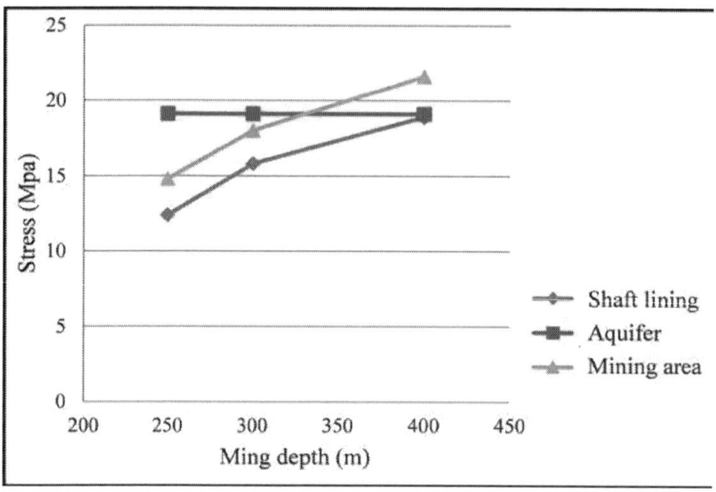

Figure 7. Impact due to mining depth (50 m to shaft).

Therefore, due to the deformation of the mining area, there is a significant stress concentration occurs in shaft lining. Particularly, a great stress concentration has occurred around the mining area and the impact range of 50 m mining width and 100 m mining width has a large difference. From above analysis, the mining width has a significant impact on the shaft lining, and the failure in shaft lining is much more severe than in the mining area.

3.5. Impact Due to Coal Seam Thickness

Three thicknesses of coal seam are set as 3 m, 6 m and 9 m for examining the impact. And the stress change in the shaft lining, aquifer and mining area is used to reveal the impact of the coal seam thickness.

Figure 11 shows the maximum principal stress in the mining area, aquifer and shaft lining. From the figure, the stress in aquifer has no obvious change. However, the stress in mining area has a significant increase with the increase of the coal seam thickness. Furthermore, the stress in mining area has outstripped the stress in shaft lining. That means the thickness change of coal seam has an impact on the mining area, but not obvious in the shaft lining and aquifer.

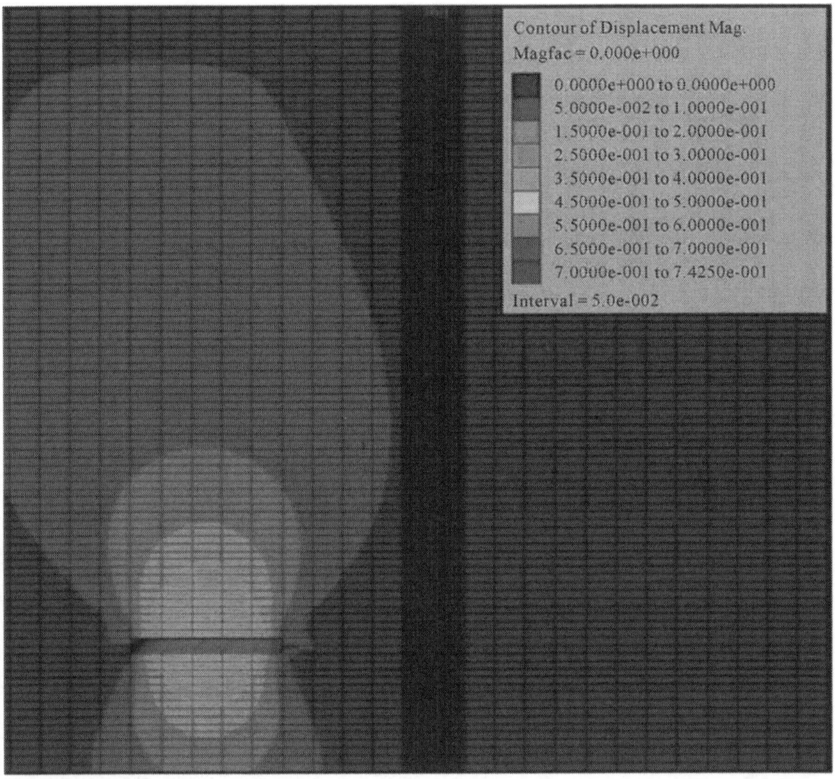

Figure 8. Vertical displacement distributions under 50 m mining width (50 m to the shaft and 400 m mining depth, 6 m thickness coal seam).

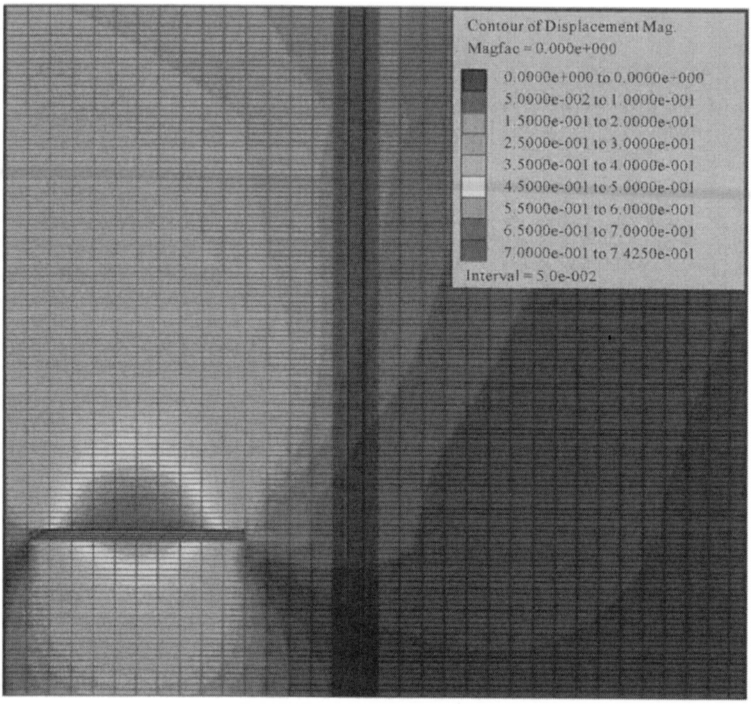

Figure 9. Vertical displacement distributions under 100 m mining width (50 m to the shaft and 400 m mining depth, 6 m thickness coal seam).

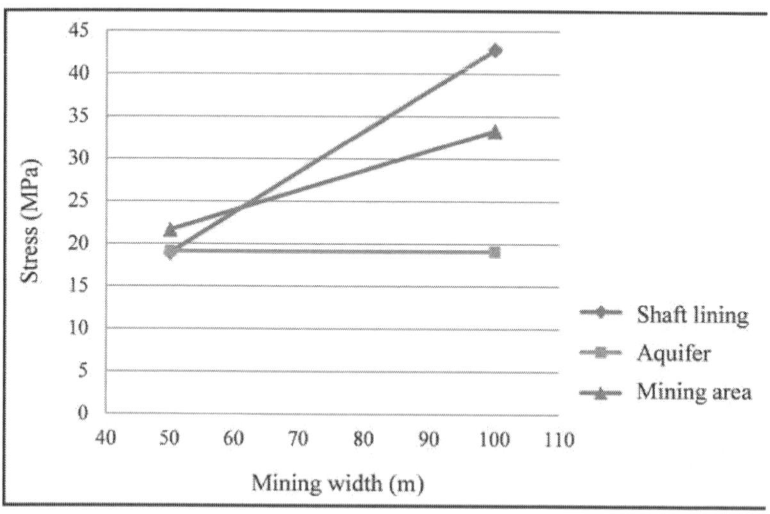

Figure 10. Impact due to mining depth due to mining width (50 m to the shaft and 400 m mining depth, 6 m thickness coal seam).

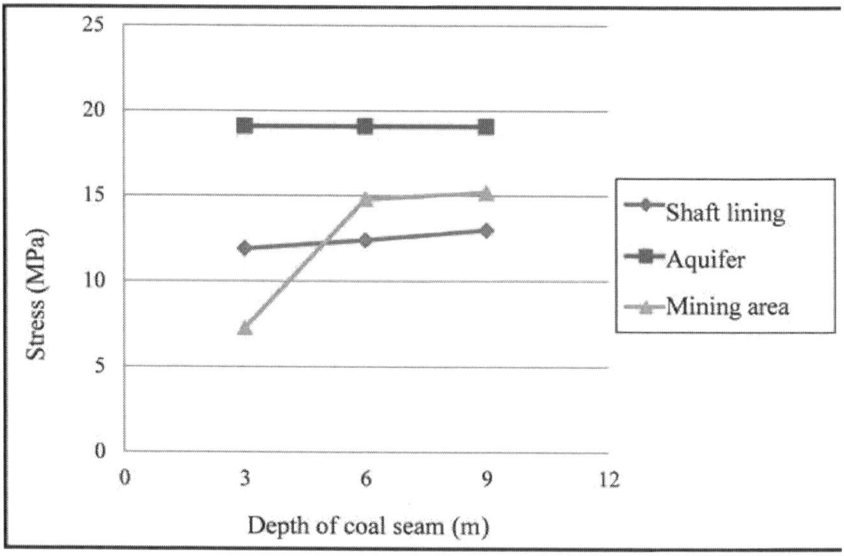

Figure 11. Impact due to coal seam thickness (50 m to shaft and 250 m mining depth).

4. CONCLUSIONS

In this study, the impact of the underground mining on the shaft lining and aquifer is analyzed by numerical simulation. The main aim is to find out the relationship between mechanisms of shaft lining rupture and the underground mining process. The impact factors such as different depth, thickness, mining width of coal seam and different distance to the shaft are used in the analysis. From a series of numerical simulations, the impact of different factors can be cleared as follows:

- The underground mining process can lead to the initiation of the failure in the soil and rock layers. If the mining location is near some certain range of the aquifer, the aquifer drainage will occur due to the fissure in aquifer connection to fissures in the bedrock. This situation should be avoided in the mining process by changing the mining location or using supporting structure.
- The impact of the mining distance to the shaft lining is obvious, if the mining location is closed to the shaft lining, the greater stress concentration will happen in shaft lining and increase the risk of occurrence of shaft lining rupture.
- When the mining location is near the shaft ling in a certain range, the increase of the mining depth also can increase the risk of occurrence of

shaft lining rupture. More seriously, the possibility of mining area collapse is raised due to the increase of mining depth.

- The mining width has a significant impact on the shaft lining, the stress in shaft lining increase fast due to the increase of mining width. Before the increase of the mining width, the support structure should be prepared to protect the mining area in case the stress transfers to shaft lining.
- The thickness change of coal seam has an impact on the mining area, but not obvious in the shaft lining and aquifer.

Therefore, when the mining area under the aquifer nears the shaft lining, there is a significant impact on the shaft lining due to mining process, and the risk of occurrence of shaft lining rupture become higher and need to be concerned.

5. ACKNOWLEDGEMENTS

Authors are grateful for financial assistance provided by the Global-Centre of Excellence in Novel Carbon Resource Science, Kyushu University.

REFERENCES

1. G. Cui and X. Cheng, "Occasions of Damaging Shaft Walls in Xuhuai District," Coal Science and Technology, Vol. 19, No. 8, 1991, pp. 46-50. [Paper reference:1]
2. L. Jing and H. Wang, "The Research about the Fracture Regularity of the Frozen Shaft Wall in Surface Soil," Journal of Huainan Institute of Technology, Vol. 20, No. 1, 2000, pp. 14-20. [Paper reference:1]
3. M. Gu, "Failure of Shaft Caused by Special Geologic Hazards and Corresponding Prevention Measures," Journal of Engineering Geology, Vol. 8, No. 2, 2000, pp. 197-201.
4. W. Yang and H. Fu, "Theoretical Investigation on Vertical Additional Force on Shaft Lining in Special Stratum," Journal of China University of Mining & Technology, Vol. 9, No. 2, 1999, pp. 129-135.
5. G. Zhou and G. Cui, "Numerical Analysis on Interaction between Shaft Wall and Surrounding Strata after Aquifer Grouting," Journal of China University of Mining & Technology, Vol. 27, No. 2, 1998, pp. 135-139.

6. H. Liu, W. Chen and Z. Wang, "Theoretical Analysis of Shaft Lining Damage Mechanism of Yanzhou Mine," Chinese Journal of Rock Mechanics and Engineering, Vol. 26, Supp. 1, 2007, pp. 2620-2626.
7. S. Bi, X. Lou and B. Xu, "On the Mechanism of Coal Mine Shaft Damage Caused by Subsidence in Xuhuai Area, Southeast China," Communications in Nonlinear Science & Numerical Simulation, Vol. 2, No. 2, 1997, pp. 75-80.
8. Yang W., Cui G. and Zhou G., "Fracture Mechanism of Shaft Lining under Special Strata Condition and the Technique Preventing the Shaft From Fracturing (Part One)," Journal of China University of Mining & Technology, Vol. 25, No. 4, 1997, pp. 1-4.
9. G. Zhou and X. Cheng, "Study on the Stress Calculation of Shaft Lining Surrounded by Special Strata," Journal of China University of Mining &Technology, Vol. 24, No. 4, 1995, pp. 24-30.
10. M. You and A. Hua, "Fracture of Rock Specimen and Decrement of Bearing Capacity in Uniaxial Compression," Chinese Journal of Rock Mechanics and Engineering, Vol. 17, No. 3, 1998, pp. 292-296. [Paper reference:1]

9

STRUCTURAL STABILITY MONITORING OF A PHYSICAL MODEL TEST ON AN UNDERGROUND CAVERN GROUP DURING DEEP EXCAVATIONS USING FBG SENSORS

Yong Li [1], Hanpeng Wang [1,*], Weishen Zhu [1], Shucai Li [1] and Jian Liu [2]

[1]Geotechnical & Structural Engineering Research Center, Shandong University, Jinan 250061, China

[2]School of Civil Engineering, Shandong University, Jinan 250061, China

ABSTRACT

Fiber Bragg Grating (FBG) sensors are comprehensively recognized as a structural stability monitoring device for all kinds of geo-materials by either embedding into or bonding onto the structural entities. The physical model in geotechnical engineering, which could accurately simulate the construction processes and the effects on the stability of underground caverns on the basis of satisfying the similarity principles, is an actual physical entity. Using a physical model test of underground caverns in Shuangjiangkou Hydropower Station, FBG sensors were used to determine how to model the small displacements of some key monitoring points in the large-scale physical model during excavation. In the process of building the test specimen, it is most successful to embed

FBG sensors in the physical model through making an opening and adding some quick-set silicon. The experimental results show that the FBG sensor has higher measuring accuracy than other conventional sensors like electrical resistance strain gages and extensometers. The experimental results are also in good agreement with the numerical simulation results. In conclusion, FBG sensors could effectively measure small displacements of monitoring points in the whole process of the physical model test. The experimental results reveal the deformation and failure characteristics of the surrounding rock mass and make some guidance for the *in situ* engineering construction.

Keywords: structural stability monitoring; FBG sensor; physical model test; underground cavern group; numerical simulation

1. INTRODUCTION

Currently, in the southwestern areas of China, numerous large-scale hydropower stations are being built, such as Ertan [1], Jinping I [2], Xiluodu [3], Dagangshan [4], Houziyan [5], and Baihetan [6]. Additionally, most of these sites require building an underground cavern group. Therefore, a large number of scientific problems urgently need to be solved regarding the stability of the underground group in the general processes of design and construction. Physical model tests, numerical modeling, and *in situ* monitoring are still three effective methods for investigating the stability of the underground cavern group. Physical model tests in geotechnical engineering could accurately simulate the excavation processes and effects on the stability of underground caverns on the basis of satisfying the similarity principles. The most important role of the physical model testing is to determine the internal stress and displacement field expected during cavern excavation. Ideally, the internal displacement results will reflect the related problems on the overall structural stability of the underground cavern group.

In the structural stability monitoring of the underground cavern group, a number of advanced measuring techniques have been utilized, such as electrical resistance strain gauges [7], Linear Variable Differential Transformers (LVDT) [8], mini multi-point extensometers [9], digital speckle photography deformation measurement (DSPD) [10], Fiber Bragg Grating (FBG) sensors [11,12,13], and others. For strain gauges bonding to the specimen for structural health monitoring, the resistive type strain gauges are always sensitive to temperature variation; therefore, it becomes necessary to account for variations in strain gauge resistance due to temperature changes. Additionally, some monitoring techniques do not respond well to instantaneous deformation associated with the cavern excavations. The main disadvantages of LVDTs are: (1) very high displacement is required for generating high voltages; (2)

shielding is required since it is sensitive to magnetic field; (3) the performance of the transducer gets affected by vibrations; and (4) it is also greatly affected by temperature changes [14]. Although the multi-point extensometer has very high measuring accuracy, it is difficult to embed it into the physical model [14]. The DSPD measuring method is also utilized in the monitoring of larger displacement fields [14].

2. PRINCIPLE AND DESIGN OF THE FBG SENSOR IN DISPLACEMENT MONITORING

2.1. Principle of the FBG Sensor

FBG sensors have been regarded as excellent transducers for a wide variety of engineering applications. One of the most commonly-used fiber optic sensors for strain and temperature measurement is the FBG sensor which was developed by Hill *et al.* in 1978 [15].They are immune to electromagnetic interference and they are small enough to be embedded into structures without causing any structural defects. The phase mask method is commonly adopted for fabrication of FBG sensors due to its high performance and simple fabrication process [16]. An FBG is composed of a refractive index "written" by the exposure to an intense UV interference pattern in the core of an optical fiber. If a broadband light is injected into the FBG sensor, it reflects the wavelength corresponding to the spacing of the different gratings, called the Bragg wavelength. The Bragg condition is expressed as the following Equation (1):

$$\lambda_B(z,t) = 2n_{eff}(z,t)\Lambda(z,t) \tag{1}$$

where λB(z, t) is the Bragg wavelength, neff(z, t) is the effective refractive index of the core mode, and Λ(z,t) is the grating period of index modulation. The wavelength, which corresponds to the Bragg condition, is reflected at the Bragg grating, and the other wavelengths pass through (see Figure 1). With this approach, the core refractive index of a bare fiber (length of 6 mm) is permanently changed after being exposed to a spatial pattern of ultraviolet light.

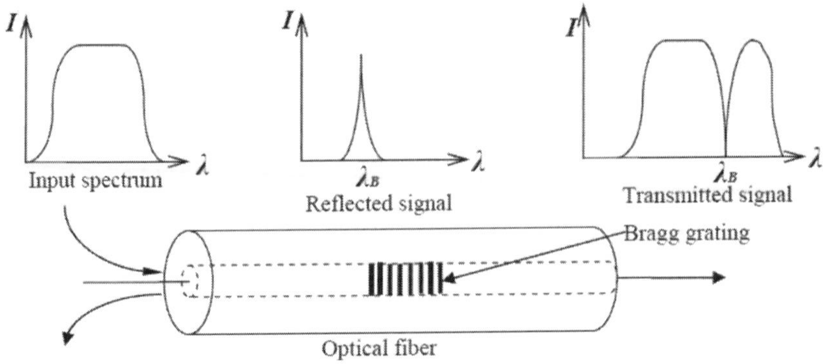

Figure 1. FBG sensor wavelength-encoding operation.

The Bragg wavelength λB(z, t) of the FBG sensor will change linearly with the applied temperature ΔTand strain ε. This relationship can be described by Kersey *et al.* [17] as below:

$$\frac{\Delta\lambda_B}{\lambda_B} = C_{\varepsilon}\varepsilon + C_T\Delta T \quad (2)$$

where λB is the original Bragg wavelength under strain free and 0 °C condition, which is the origial condition and the other conditions could be calculated based on it; ΔλB is the change in the Bragg wavelength due to the variation of strain and temperature; Cε and CT are the calibration coefficients of strain and temperature. The typical values of Cε and CT are approximately 0.78 × 10−6 με−1 and 6.7 × 10−6 °C−1 respectively [18,19]. The above relationship could be expressed more specifically as below:

$$\frac{\Delta\lambda_B(z,t)}{\lambda_B(z,t)} = (1-p_e)\cdot\varepsilon(z,t) + (\alpha+\xi)\cdot\Delta T(z,t) \approx 0.78\varepsilon(z,t) + 6.7\times10^{-6}\Delta T(z,t) \quad (3)$$

where pe is the elastic optical coefficient; α and ξ are the coefficients of temperature effect; ΔT(z, t) is change of temperature.

2.2. Design and Calibration of the FBG Sensing Bar

The FBG sensing bars are developed based on the deformations of an elastic axiasymmetric beam as shown in Figure 2. The length and radius of the beam are L and R, respectively. This beam is fixed at one end and subjected to arbitrary transverse and/or axial loading. An imagined plane H passes through the beam perpendicular to z axis. The beam axis on this cross section has deflections u and v in the x and ydirections and a tension (or compression) w in the z direction under the effect of load combinations. According to Euler-Bernoulli beam theory [20], the distributions of strain ε(z) is associated with the distributions of normal axial force FN(z) and the bending moment M(z) along the neutral line of the beam, which can be expressed as below:

$$\varepsilon_z = \frac{F_N(z)}{EA} \pm \frac{M_x(z)\cdot R}{EI_x} \pm \frac{M_y(z)\cdot R}{EI_y} \tag{4}$$
$$= \varepsilon_A(z) \pm \varepsilon_{Tx}(z) \pm \varepsilon_{Ty}(z)$$

where, εA(z) is the strains induced by the normal axial force, εTx(z) and εTy(z) are the strains induced by the transverse loading in the x and y directions, which are associated with the bending moment Mx(z) and My(z), respectively. E is the Young's modulus; A is the cross sectional area; Ix and Iy represent the moments of inertia with respect to the *x* and *y* axes of the beam, respectively.

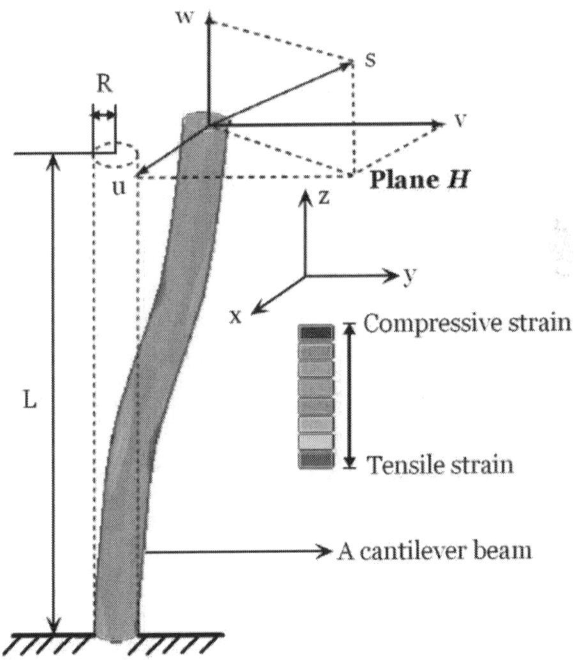

Figure 2. A schematic illustration of a cantilever beam under arbitrary transverse and/or axial loading.

The lateral deflections in the x and y directions and the tensile (or compressive) displacement in the $z$direction can be calculated by integration of strains as the following:

$$\begin{cases} u = \dfrac{1}{R}\iint \varepsilon_{Tx}(\mathrm{z})dzdz \\ v = \dfrac{1}{R}\iint \varepsilon_{Ty}(\mathrm{z})dzdz \\ w = \int \varepsilon_{A}(\mathrm{z})\, dz \end{cases} \tag{5}$$

The displacements of the three directions could be calculated by Equation (5) and the specified boundary conditions. As for the cantilever beam shown in Figure 2, the boundary at the fixed end has the relationships as below:

$$\begin{cases} u_0 = v_0 = w_0 = 0 \\ \dfrac{du}{dz} = 0 \\ \dfrac{dv}{dz} = 0 \end{cases} \tag{6}$$

Based on the beam theories above, the FBG sensing bar (see Figure 3) is designed and manufactured from a grooved plastic or rubber bar with a diameter of 10 mm. This type of plastic or rubber bar has sufficient elasticity, which could satisfy the cooperation deformation with the surrounding rock mass in the physical model. Four optical fibers are adhered in the grooves and covered with epoxy resin. Each optical fiber contains a series of FBG strain sensors at regular intervals. The length of the bar and the spacing of FBG sensors are changeable according to different experimental conditions.

The FBG sensing bar is designed to be pre-embedded vertically in the process of the construction of a physical model or be inserted into a borehole, working like a cantilever beam under arbitrary axial and/or transverse loading. When the excavations of the cavern group are performed, the FBG sensing bar is subjected to bending and tension (or compression). The multiple FBG sensors measure the strain distributions along the bar perpendicularly. In the plane H in Figure 2, the strains induced by axial and/or transverse loading could be calculated by:

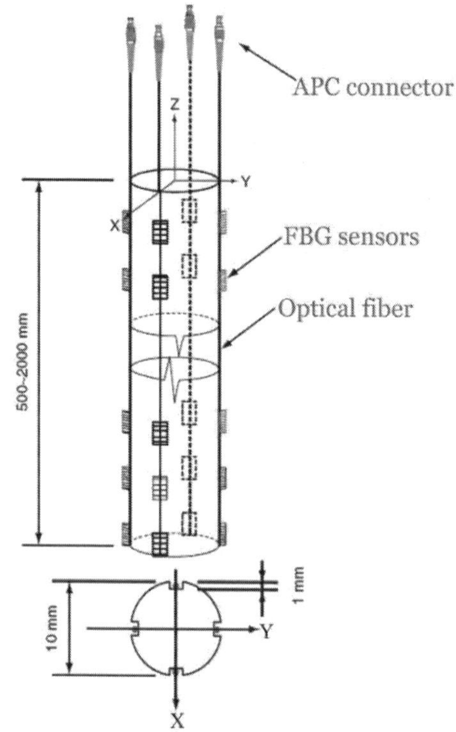

Figure 3. A Design of the FBG sensing bar for 3-D displacement measurement.

$$\begin{cases} \varepsilon_A^i = \dfrac{1}{4}(\varepsilon_{ia} + \varepsilon_{ib} + \varepsilon_{ic} + \varepsilon_{id}) \\ \varepsilon_{Tx}^i = \dfrac{1}{2}(\varepsilon_{ia} - \varepsilon_{ic}) \\ \varepsilon_{Ty}^i = \dfrac{1}{2}(\varepsilon_{ib} - \varepsilon_{id}) \end{cases} \tag{7}$$

where εia, εib, εic, and εid are the strains measured by the four FBG sensors on the plane H, respectively. Combining Equations (5) and (7), and performing linear interpolation of strain distributions, the distributions of internal displacements in three dimensions could be obtained.

To verify the effectiveness of the FBG sensing bar, the authors performed a calibration test (tension and deflection) using both the FBG sensing bar and LVDT sensors as shown in Figure 4. During the test, different loads were imposed to the FBG sensing bars. Meanwhile, the deflection values were obtained by using FBG and LVDT sensors. Figure 5 shows the comparative curves of the deflection values obtained by FBG and LVDT sensors.

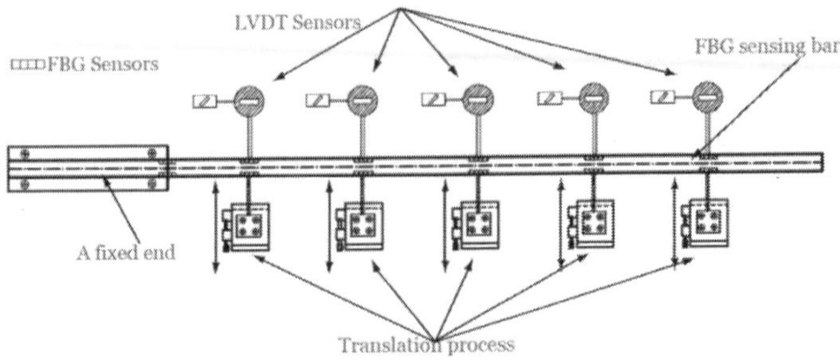

Figure 4. Installment of FBG and LVDT sensors in the deflection test of FBG sensing bar.

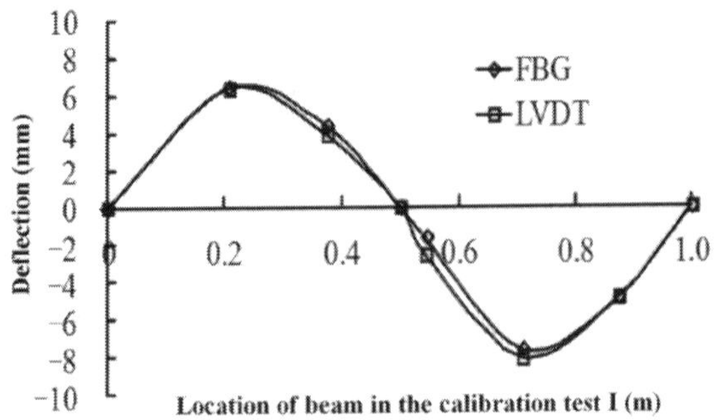

(a)

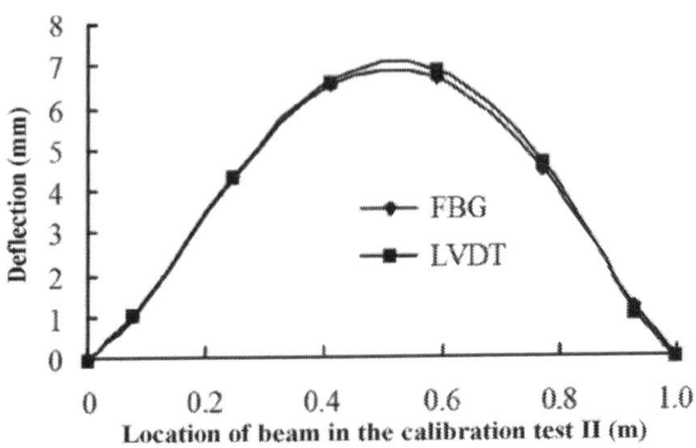

(b)

Figure 5. Comparative curves of the deflection values obtained by FBG and LVDT sensors. (**a**) The calibration test of tension; (**b**) The calibration test of deflection.

The calibration tests results show that the FBG sensing bar could easily obtain the applied deflection values along the bar. If the FBG sensing bar is subjected to a lateral displacement at the free end, the relationship between the applied displacement and the maximum strain should be $D\varepsilon max=L23R$. In this deflection calibration test, the resolution of FBG sensing bar for measuring deflection could be calculated as $12/(3\times5\times10-3)\times1\times10-6=6.7\times10-5m=67\ \mu m$, and the maximum measuring deflection for the FBG sensing bar could be obtained as $12/(3\times5\times10-3)\times3000\times10-6=0.2\ m=200\ mm$. From the above comparative curves, it is concluded that the deflection measured by the FBG sensors is in good agreement with those obtained by LVDT sensors. Moreover, the FBG sensing bar is an appropriate tool for measuring internal displacement in the physical model test.

3. INTRODUCTION TO THE PHYSICAL MODEL TEST

3.1. Project Description

The Shuangjiangkou Hydropower station is located on the Dadu River in Sichuan Province, China. The river meanders in an entrenched valley with a wall height to 1000 m and slope between 35° and 60°. Below an altitude of 2800 m the valley is a near-symmetrical V-shape. *In situ* stress fields near the underground cavern complex are strongly influenced by the incised terrain, active tectonics, and the high rate of incision and corresponding unloading at the site. *In situ* stresses reach 38 MPa at a depth of approximately 600 m. The rock mass is composed of medium to fine-grained granites with no apparent foliation.

The underground power-house contains four turbines with a total capacity of 2 GW. The underground cavern group consists of the main powerhouse, the transformer house, and the surge chamber (see Figure 6). The axial direction of the cavern complex is N100W. From observations at other large to medium-sized underground hydropower stations such as Ertan, Xiluodu, Xiaowan, Pubugou, and Jinping hydropower stations, the spacing between the three main caverns was determined from a preliminary stability analysis [1]. The result was to separate the main powerhouse and the transformer house by a 45 m wide pillar, and to separate the transformer house and the surge chamber by 40 m. For the main powerhouse, the transformer house, and the surge chamber, the heights are 67.05 m, 26.5 m, and 80.2 m, respectively. Their spans are 28.3 m, 18 m, and 20 m, respectively.

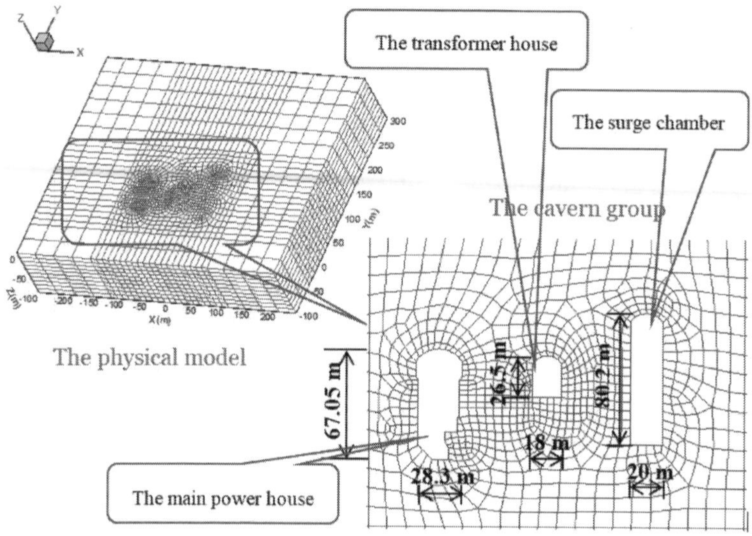

Figure 6. The layout of the physical model with the cavern group.

3.2. Three Dimensional Physical Model Test of an Underground Cavern Group

Physical model tests were conducted to investigate the stability of the cavern group, which contains three openings, all subjected to high *in situ* stresses. The scaling ratio between the physical model and the prototype is 1/200.

3.2.1. Steel Structure Frame for the Physical Model Test

The first step of the physical model test is to design and manufacture a steel structural frame. This steel structure guarantees that the physical model test is in a true three-dimensional state. This structure has technical advantages such as high stiffness, great stability and flexibility of assembly, and easy adjustment of its dimensions. Figure 7 shows the whole true 3D steel frame.

3.2.2. Analogous Material for Physical Model Test

A new type of rock analogy material has been developed in the physical model test, which is made from iron mineral powder, barite powder, quartz powder, and alcoholic solution with rosin. The iron mineral powder, barite powder, and quartz powder are skeletal materials while the alcoholic solution with rosin is a cementing agent. Table 1 shows the physico-mechanical parameters of the rock mass and analogy material.

(a) (b)

Figure 7. The 3D steel frame structure for the physical model test. (**a**) A front view; (**b**) A side view.

Table 1. The physico-mechanical parameters of the prototype and analogy material.

Type of Material	Density (KN/m^3)	Young's Modulus (MPa)	Cohesion (MPa)	Internal Friction Angles (°)	Compressive Strength (MPa)	Poisson's Ratio
Rock Mass	26.5	3000	2	40.36	80	0.2
Analog Material	26.5	15	0.01	40.36	0.4	0.2

3.2.3. The Installation of FBG Sensing Bar in the Physical Model

During the construction of the model, three holes were prepared for installation of FBGs near the openings of the main powerhouse and the surge chamber. The holes were 2000 mm in length and 18 mm in diameter. The center of the hole was 19 mm away from the side wall of the cavern, as shown in Figure 8. The sensing bars were embedded in the model vertically, in order to satisfy the deformation compatibility between the bar and the model. Initially, the ordinary rubber, latex, and silica gel were selected to be the bonding material. After a large number of trial tests, the ordinary silica gel was selected as the bonding material. The sensing bar was inserted into the hole immediately after a reasonable amount of silica gel was injected into the hole as the grouting material. The installation shall be as vertical and smooth as possible so as to avoid axial rotation of the sensing bar. In addition, the direction of the

FBGs adhered on the bar shall be perpendicular to the axial direction of the cavern. In this way, the measurement results can truly reflect the deformation of caverns. After 24-hour curing of silica gel, the bonding between sensors and the model was tested to be strong enough, and satisfy the deformation compatibility condition.

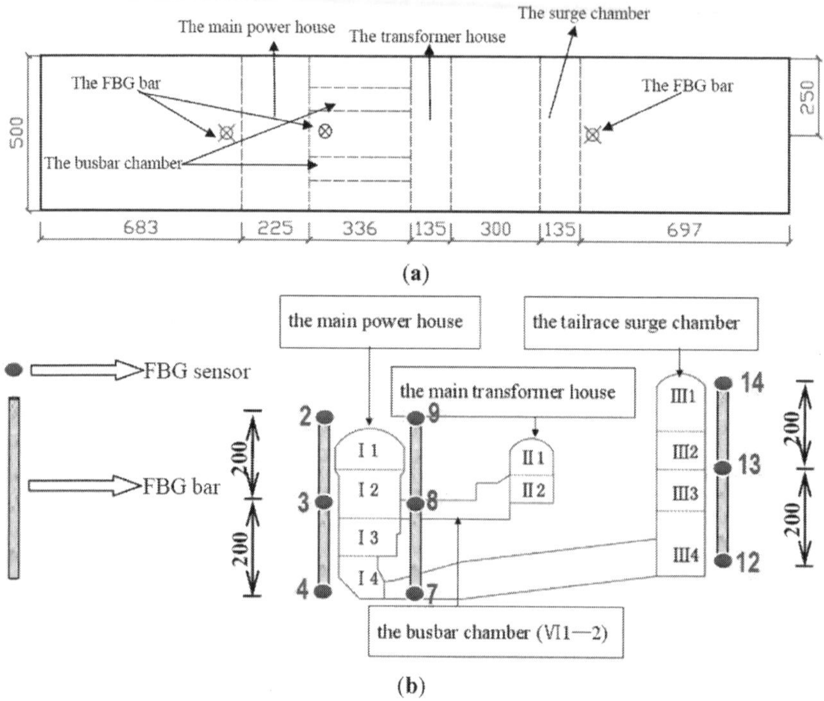

Figure 8. Layout of the FBG sensing bars (Unit: mm). (**a**) A top view; (**b**) A front view.

3.2.4. Excavations of the Cavern Group and Structural Stability Monitoring of Surrounding Rock Mass

The use of underground caverns in hydropower stations is increasing. However, excavation of such spaces results in a change in stress distribution; the changes alter the mechanical properties of rock mass, such as strength and deformability. In particular, the displacements of critical points at the arch crown and side walls of the caverns are the vital factor to the longevity of the cavern group. If extensive deformation occurs around the caverns, rockburst and failure may occur. Therefore, it is critical to monitor the internal deformation of the underground caverns during excavation in real time.

Due to the space restriction of the model, twenty excavation subsequences in the field were simplified to ten subsequences in the model test (see Figure 8). They are I1, II1, III1, I2, II2, III2, I3, III3, I4, and III4, respectively. The drilling and blasting method was adopted in the *in situ* project, but this method is difficult to be performed in the physical model test. Therefore, the caverns were excavated by special drilling tools with sharp heads. The excavation footage of every substep is 5 cm. The whole excavation subsequences can be divided into 10 circles, and the last two substeps are to excavate the busbar chambers and the other openings, so there are 102 steps in total. During the excavation, the pre-installed FBG sensors took monitoring data automatically. When a circle was completed, the column bolts pre-embedded for bolt holes were pulled out, the rock bolts were placed, and grouting was applied (the column bolts are the same size as the rock bolts, made of slim iron rods). All the pre-installed FBG sensors recorded the data during the excavation.

4. NUMERICAL SIMULATION OF THE PHYSICAL MODEL TEST AND COMPARATIVE ANALYSIS WITH THE MONITORING RESULTS USING FBG SENSING BARS

4.1. Numerical Simulation of the Physical Model Test

In the numerical simulations, a 3D finite element model, which had the same dimensions as the physical model, was built. The shotcrete layer and the surrounding rock masses were represented by elements with only axial stiffness taken into account. The model was divided into 39,798 nodes and 35,440 elements, as shown in Figure 9. The surrounding rock masses of the underground cavern group were of good integrity, joints did not develop, and seepage of water was not observed in the initial exploration audits. These factors were, therefore, not considered in the numerical simulations. The Drucker-Prager criterion was adopted and the material properties were obtained according to *in situ* physico-mechanical parameters as shown in Table 1. Stability analysis was performed by using the finite-difference method and the FLAC^{3D} code.

4.2. A Comparative Analysis between the Monitoring Results and the Numerical Results

During the excavation process, the FBG monitoring results show that the maximum displacements appear at the middle part of the upstream side wall of the mainpower house and downstream side wall of the surge chamber (*i.e.*, FBG #3 and #13). The maximum displacements are 0.182

mm and 0.257 mm, respectively. It is reasonable that the height of the surge chamber is greater than that of the main powerhouse.

Figure 10 and Figure 11 show the horizontal displacement curves of FBG #3 and #13. The horizontal axis represents the excavations steps. It is shown that the monitoring results agree well with the numerical results. As the excavations continue, the deformation of each point is increasing, and ultimately tends toward a stable value.

There was no brittle failure or rockbursts observed during the excavation process. It is concluded that the excavation scheme is feasible and the supporting scheme has great impacts on the reinforcement of the surrounding rock mass.

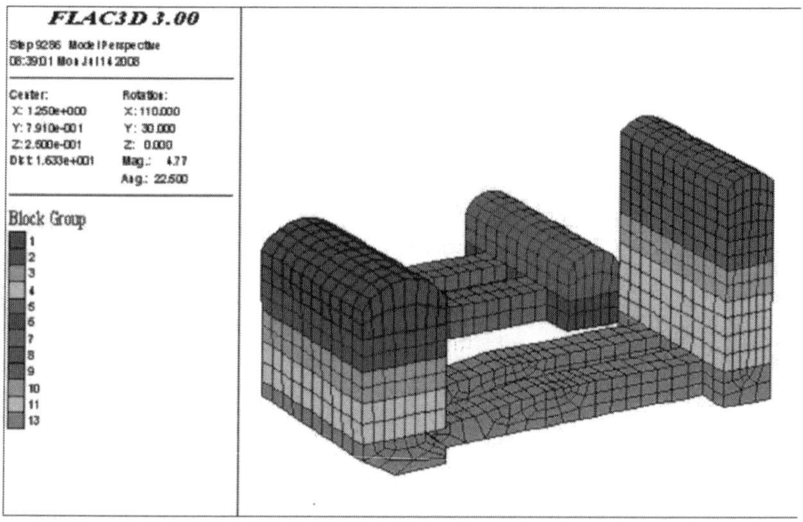

Figure 9. A 3D numerical model of the underground cavern group.

5. SUMMARY AND CONCLUSIONS

This paper presents the development of a novel FBG sensing bar for monitoring internal displacements of a large-scale physical model test on an underground cavern group. The FBG sensing bar is designed to be embedded in the physical model and measure the internal displacements accurately and automatically. The main conclusions of this work are:

(1) It is feasible to use FBG sensors in the internal displacement monitoring of a large-scale physical model test. The FBG sensor has more advantages than other conventional sensors, such as its small size, high

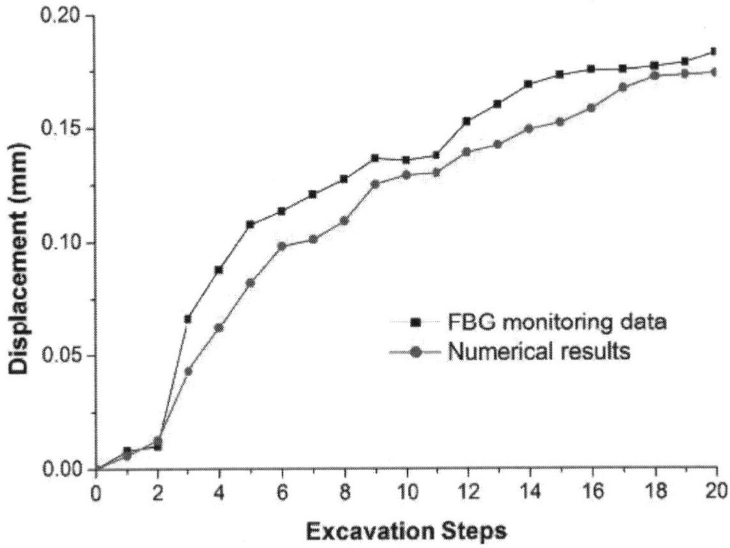

Figure 10. The horizontal displacement curves with the excavation steps obtained by FBG sensing bar and numerical simulation at monitoring point #3.

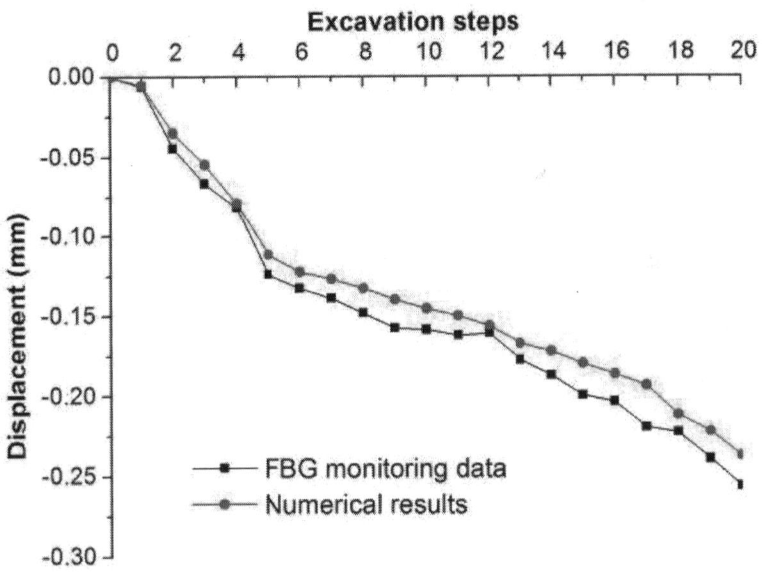

Figure 11. The horizontal displacement curves with the excavation steps obtained by FBG sensing bar and numerical simulation at monitoring point #13.

measuring accuracy, high sensitivity, ability to be embedded in the physical model, strong anti-interference ability, wide measurement range, and online continuous detecting ability.

(2) The design and installation measures of FBG sensing bars in the physical model is proved to be successful and worth popularizing.

(3) In the process of the cavern group excavations, the displacements of the surrounding rock mass of the cavern group continue to increase as the excavation go on, but the surrounding rock mass is always in a stable state. In this whole stage, the displacement monitoring results obtained by FBG sensors are in good agreement with those obtained by numerical results.

ACKNOWLEDGMENTS

The work was supported by the National Science and Technology Support Program of China (2015BAB07B05), the Natural Science Foundation of Shandong Province (BS2012NJ006) and the Specialized Research Fund for the Doctoral Program for Higher Education (No. 20110131120034). We would like to show our great acknowledgments to Yin J.H. from The Hong Kong Polytechnic University. We would also like to express our sincere gratitude to the editor and the two anonymous reviewers for their valuable contributions to this paper.

AUTHOR CONTRIBUTIONS

This present research article is based on the research work of Y.L. and H.W. During the research work, W.Z., S.L. and J.L. have provided substantial contributions to the theoretical innovation and experimental guidance. Y.L. performed a lot of work in laboratory experiments and H.W. conducted the numerical simulation and analyzed the related data. Finally, Y.L. finished the whole writing of this paper.

CONFLICTS OF INTEREST

The authors of the paper declare that there is no conflict of interest regarding the publican of this paper. The authors do not have a direct financial relation with the commercial identity that might lead to a conflict of interest for any of the authors.

REFERENCES

1. Zhu, W.S.; Sun, A.H.; Sui, B. Systematic analysis of underground cavern groups in background for Ertan project. Chin. J. Undergr. Sp. Eng. 2005, 1, 15–18.

2. Li, Y.; Zhu, W.S.; Fu, J.W.; Guo, Y.H.; Qi, Y.P. A damage rheology model applied to analysis of splitting failure in underground caverns of Jinping I hydropower station. Int. J. Rock Mech. Min. Sci.2014, 71, 224–234.

3. Lu, Y.M.; Fan, Q.X.; Zhou, S.W.; Li, B.F. Key technologies for construction of Xiluodu high arch dam on Jinsha River. J. Hydr. Eng. 2013, 32, 187–195.

4. Yang, W.D.; Zhang, Q.Y.; Li, S.C.; Wang, S.G. Estimation of in situ viscoelastic parameters of a weak rock layer by time-dependent plate-loading tests. Int. J. Rock Mech. Min. Sci. 2014, 66, 169–176.

5. Xu, N.W.; Li, T.B.; Dai, F.; Li, B.; Zhu, Y.G.; Yang, D.S. Microseismic monitoring and stability evaluation for the large scale underground caverns at the Houziyan hydropower station in Southwest China. Eng. Geol. 2015, 188, 48–67.

6. Jiang, Q.; Feng, X.T.; Hatzor, Y.H.; Hao, X.J.; Li, S.J. Mechanical anisotropy of columnar jointed basalts: An example from the Baihetan hydropower station, China. Eng. Geol. 2014, 175, 35–45.

7. Zhang, J.; Xia, Q.; Cheng, Y.Y.; Wu, Z.S. Strain flexibility identification of bridges from long-gauge strain measurements. Mech. Syst. Signal Process. 2015, 62, 272–283.

8. Park, Y.T.; Shin, F.K.; Jang, W.B.; Jeong, Y.H.; Jang, S.M. The analysis on the performance characteristics for design of a linear variable differential transformer (LVDT). Els. Appl. Elect. Mat.1995, 6, 345–348.

9. Zhu, W.S.; Li, Y.; Li, S.C.; Wang, S.G.; Zhang, Q.B. Quasi-three-dimensional physical model tests on a cavern complex under high in-situ stresses. Int. J. Rock Mech. Min. Sci. 2011, 48, 199–209.

10. Chiang, F.P. Super-resolution digital speckle photography for micro/nano measurements. Opt. Laser Eng. 2009, 47, 274–279.

11. Kinet, D.; Mégret, P.; Goossen, K.W.; Qiu, L.; Heider, D.; Caucheteur, C. Fiber bragg grating Sensors toward structural health monitoring in composite materials: Challenges and solutions. Sensors 2014,14, 7394–7419.

12. Gage, J.R.; Wang, H.F.; Fratta, D.O.; Turner, A.L. In situ measurements of rock mass deformability using fiber bragg grating strain gauges. Int. J. Rock Mech. Min. Sci. 2014, 71, 350–361.

13. Gage, J.R.; Fratta, D.O.; Turner, A.L.; MacLaughlin, M.M.; Wang, H.F. Validation and implementation of a new method for monitoring in situ strain and temperature in rock masses using fiber-optically instrumented rock strain and temperature strips. Int. J. Rock Mech. Min. Sci. 2013, 61, 244–255.

14. Ribeiroa, D.; Calçadab, R.; Ferreirac, J.; Martinsc, T. Non-contact measurement of the dynamic displacement of railway bridges using an advanced video-based system. Eng. Struct. 2014, 75, 164–180.

15. Hill, K.O.; Fujii, Y.; Johnson, D.C.; Kawasaki, B.S. Photosensitivity in optical fiber waveguides: Application to reflection filter fabrication. Appl. Phys. Lett. 1978, 32, 647–649.

16. Hill, K.O.; Meltz, G. Fiber Bragg grating technology fundamentals and overview. J. Lightwave Technol.1997, 8, 1263–1276.

17. Kersey, A.D.; Davis, M.A.; Patrick, H.J.; LeBlanc, M.; Koo, K.P.; Askins, C.G.; Putnam, M.A.; Friebele, E.J. Fiber grating sensors. J. Lightwave Technol. 1997, 15, 1442–1463.

18. Zhu, H.H.; Yin, J.H.; Zhang, L.; Jin, W.; Dong, J.H. Monitoring internal displacements of a model dam using FBG sensing bars. Adv. Struct. Eng. 2010, 13, 249–261.

19. Xu, D.S.; Yin, J.H.; Cao, Z.Z.; Wang, Y.L.; Zhu, H.H.; Pei, H.F. A new flexible FBG sensing beam for measuring dynamic lateral displacements of soil in a shaking table test. Measurement 2013, 46, 200–209.

20. Labuschagnea, A.; van Rensburga, N.F.J.; van der Merweb, A.J. Comparison of linear beam theories.Math. Comput. Model. 2009, 49, 20–30.

10

EVALUATION OF INVERSE MODELING TECHNIQUES FOR PINPOINTING WATER LEAKAGES AT BUILDING CONSTRUCTIONS

A. W. M. van Schijndel

Unit Building Physics and Services, Department of the Built Environment, Eindhoven University of Technology, Eindhoven, Netherlands

ABSTRACT

The location and nature of the moisture leakages are sometimes difficult to detect. Moreover, the relation between observed inside surface moisture patterns and where the moisture enters the construction is often not clear. The objective of this paper is to investigate inverse modeling techniques as a tool for the detection of moisture leakage locations in building constructions from inside surface moisture patterns. It is concluded that although the presented methodology is promising, more research is needed to confirm its usability.

Keywords: Inverse Modeling, Building, Construction, Moisture, Leakage

1. INTRODUCTION

Hunting Lodge St. Hubertus is one of the most prominent buildings from the beginning of the twentieth century and is noted in the top 100 list of Dutch monuments. The conservation of the building and its interior are of great importance. The Dutch Government Building Department, which takes care of the maintenance of the building, has expressed their concern about the observed damage due to high moisture levels by the rain that finds its way to the interior at places of inadequate detailing and therefore causes damage mainly near openings in the facade and on the inside of the facade below balconies. The main problem is that the location and nature of the moisture leakages are not easily detectable. We often don't know the relation between the observed inside surface moisture patterns and where the moisture enters the construction. The objective is to investigate inverse modeling techniques as a tool for the detection of moisture leakage locations in building constructions, i.e. we want to investigate the (in)possibilities of pinpointing moisture leakages from inside surface moisture patterns using inverse modeling techniques.

1.1. Summary of the Observed Moisture Problems at the Hunting Lodge St. Hubertus

Hunting Lodge St. Hubertus is located on the northern side of the Dutch National Park "De Hoge Veluwe". The Hunting Lodge is built as a guesthouse between 1916 and 1922, by Holland's most well-known architect from that time, H. P. Berlage. The building consists of a low-rise rectangular volume with wings that stretch out diagonally and with a characteristic high tower of over 30 meters height in the middle of the building (see Figure 1). A large pond is situated south-west of the building and the building is surrounded by forest in all other directions.

The damage that occurs in the tower was systematically inspected to enable a thorough assessment of the possible causes of the moisture problems by Briggen et al. (2009). The damage on the inside of the tower, and where possible also on the outside, is systematically inspected. The location and type of each moisture problem are documented in a table, illustrated with a picture of the damage. The moisture problems that manifest themselves in the tower of the Hunting Lodge can be divided in the following categories: efflorescence, cracking, soiling, moist spots, mechanical damage and biological growth.

A few pictures of the moisture damage that occurs in the tower are shown in Figure 2 (Briggen et al. [1]).

Regarding the location of the damage it can be concluded from the inspection that most damage occurs on the interior surface of the south-west facade of the tower. Since the prevailing wind direction in the

Netherlands is south-west, which means that the south-west facade of the tower is subjected to wind-driven rain the most, there appears to be a connection between the rain load of the facade and the damage on the inside. There are no clear differences between the damage on lower or higher floors or between the damage on the middle and on the sides of the facade. Most damage occurs near openings in the facade and on the interior surface of the facade below balconies.

Figure 1. Hunting lodge St. Hubertus.

Figure 2. Observed moisture damage in the tower of the building: moist spots and efflorescence.

1.2. Methodology

The research method contained the following steps:

(Step 1) Measurements of the external climate, indoor climate and surface conditions (Section 2).

(Step 2) Computational model development and calibration of the parameters (Section 3).

(Step 3) From one of the prominent observed moisture spot, an inverse modeling technique was used to determine the moisture entrance for pinpointing the most likely water leakage (Section 4).

(Step 4) Discussion and conclusions of the approach (Section 5).

2. MEASUREMENTS

The data set is part of the measurement program at the Hunting Lodge St. Hubertus site, performed during 2006- 2007 by Briggen [1] . Details of this project can be found in Briggen et al. [2] [3] . One of problems seemed to be high moisture contents at the inside surface of the facade of the tower. The construction of this facade is shown in Figure 3.

The outside climate conditions were measured by a weather station within 50m from the building. The inside air temperature and relative humidity were measured using standard equipment (seeFigure 4). A representation of inside surface conditions was obtained by placing a small box (5 cm × 5 cm × 1 cm) against the wall and measures the air temperature and relative humidity inside. The estimation of the measurement error of this method is left over for future research.

The data consists of the measured time series of the indoor and outdoor climate as presented inFigure 5 and Figure 6.

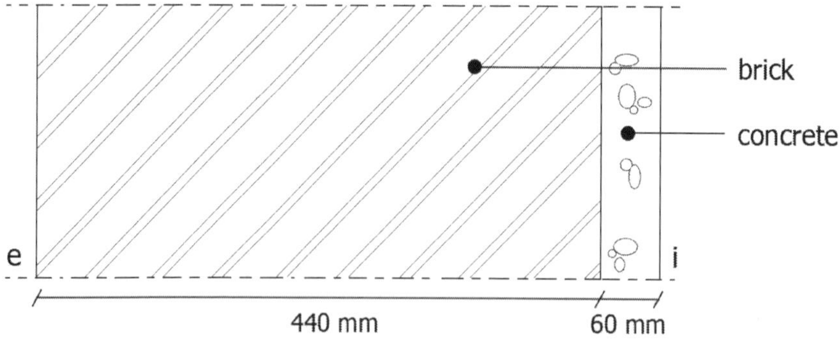

Figure 3. The building facade.

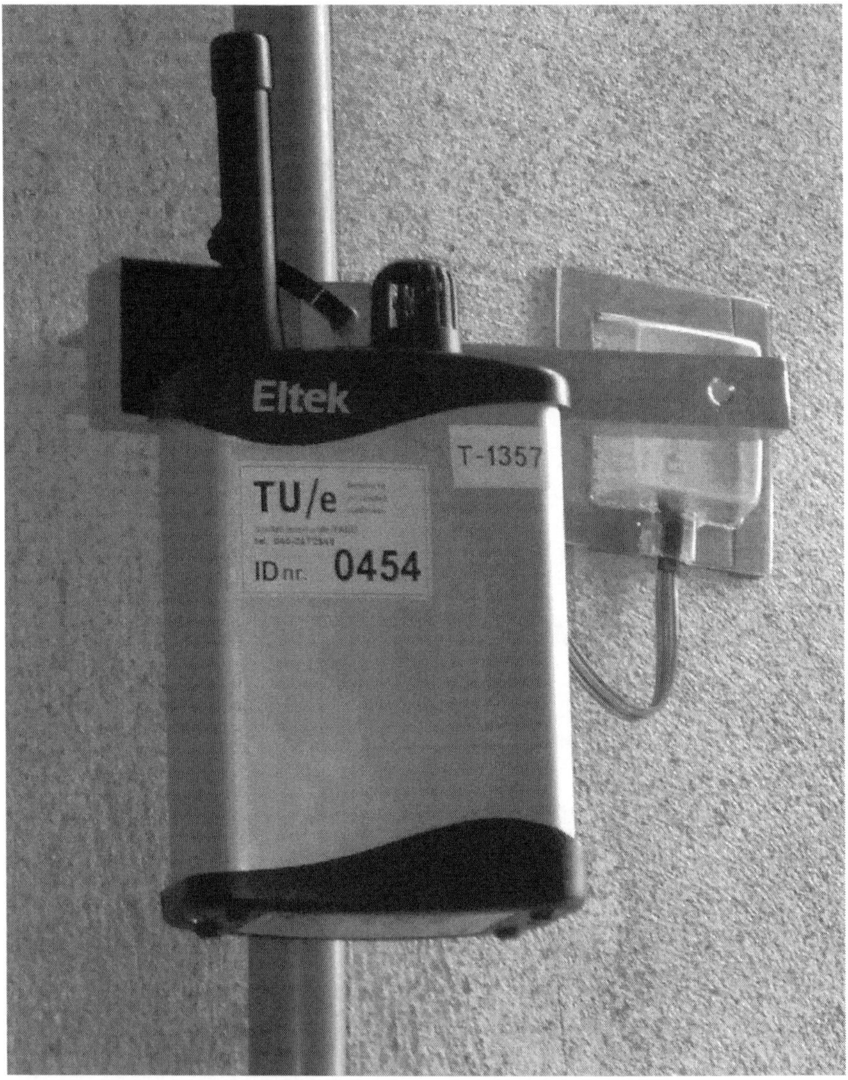

Figure 4. Measurement of the surface temperature and relative humidity using a box.

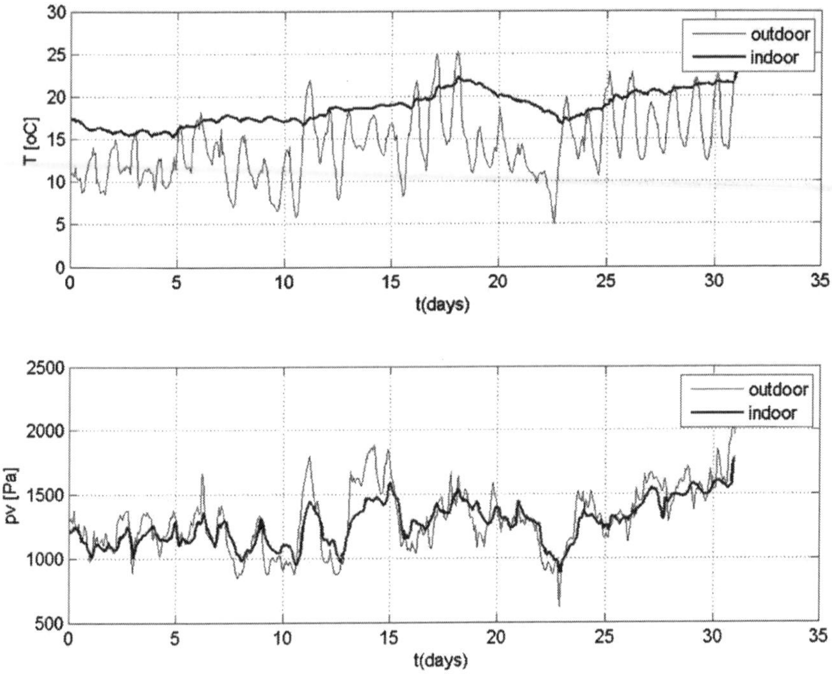

Figure 5. The measured air temperatures (top) and calculated vapour pressures (bottom, from measured T/RH).

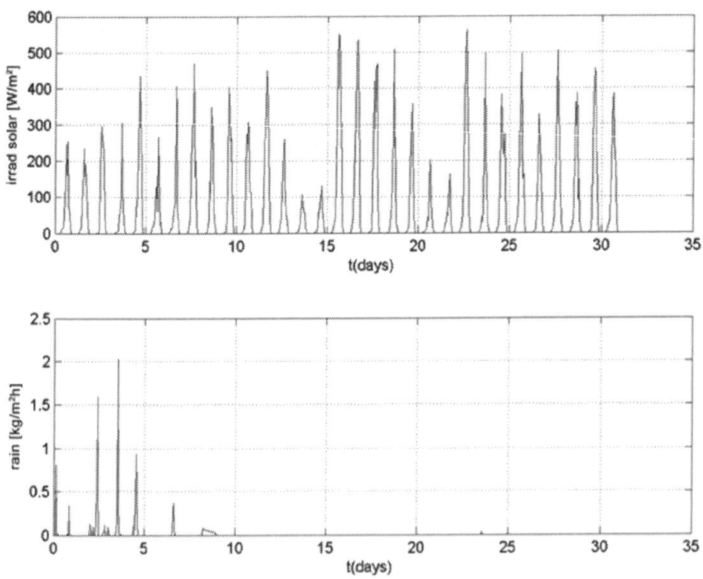

Figure 6. The measured solar irradiance (top) and rain intensity (bottom).

3. MODELING

The multiphysics modeling approach of van Schijndel [4] - [6] is used. The heat and moisture transport can be described by the following PDEs:

$$C_T \frac{\partial T}{\partial t} = \nabla \cdot \left(K_{11} \nabla T + K_{12} \nabla LPc\right) \tag{1}$$

$$C_{LPc} \frac{\partial LPc}{\partial t} = \nabla \cdot \left(K_{21} \nabla T + K_{22} \nabla LPc\right) \tag{2}$$

where t is time [s]; T is temperature [°C]; The reminder of the terms in the heat (1) and moisture Equation (2) are explained below:

$$LPc = {}^{10}\log\left(Pc\right) \tag{3}$$

$$C_T = \rho \cdot c \tag{4}$$

$$K_{11} = \lambda \tag{5}$$

$$K_{12} = -l_{lv} \cdot \delta_p \cdot \phi \cdot \frac{\partial Pc}{\partial LPc} \cdot \mathrm{Psat} \cdot \frac{M_w}{\rho_a RT} \tag{6}$$

$$C_{LPc} = \frac{\partial w}{\partial Pc} \cdot \frac{\partial Pc}{\partial LPc} \tag{7}$$

$$K_{22} = -K \cdot \frac{\partial Pc}{\partial LPc} - \delta_p \cdot \phi \cdot \frac{\partial Pc}{\partial LPc} \cdot \mathrm{Psat} \cdot \frac{M_w}{\rho_a RT} \tag{8}$$

$$K_{21} = \delta_p \cdot \phi \cdot \frac{\partial \mathrm{Psat}}{\partial T} \tag{9}$$

where Pc is capillary pressure [Pa]; ρ is material density [kg/m³]; c is specific heat capacity [J/kg·K]; λ is thermal conductivity [W/mK]; l_{lv} is specific latent heat of evaporation [J/kg]; δ_p vapour permeability [s]; ϕ is relative humidity [-]; Psat is saturation pressure [Pa]; $M_w = 0.018$

[kg/mol]; $R=8.314$ [J/mol·K]; ρ_a is air density [kg/m³]; w is moisture content [kg/m³]; K is liquid water permeability [s].

MatLab is used for the implementation of material and boundary properties. These functions are used to convert measurable material properties such as K, ϕ, δ_p and λ which are dependent on the moisture content into PDE coefficients which are dependent on the LPc and T. This is schematically shown in Figure 7.

The material database of DELPHIN [7] is used to provide material properties for the first guess. For brick, the Brick material properties of DELPHIN are used with constant $\rho=1700$; $c=840$; $\lambda=0.85$ and variable moisture properties using the tables. For concrete, the Lime plaster properties of DELPHIN $(\rho=1800;\ c=840;\ \lambda=1.05)$ are used in the same way. From these data, the PDE coefficients were determined together with the boundary conditions implemented using the COMSOL [8] model of Section 3. Figure 8 and Figure 9 show the results.

Figure 8 shows that the simulated inside surface temperature is already quite close to the measured one.

The simulated relative humidity at the inside surface of Figure 9 seems to be less close to the measured one compared to the previous figure. This gives also rise to the just mentioned questions. For each material and at each point the vapour pressure can be calculated using a similar corresponding function.

4. DETERMINATION OF MOISTURE SOURCE CHARACTERISTICS

In this section we try to reproduce the following observed moisture spots (see Figure 10).

The modeling approach of the previous section was used. The mesh of (simplified) geometry is presented in Figure 11.

The first step of the inverse modeling procedure is to switch one or more boundary conditions from dry into wet and then investigate it's effect on the inside surface moisture print. For example,Figure 13 shows the simulated profile at the inner surface by switching the location provided inFigure 12 from dry into wet.

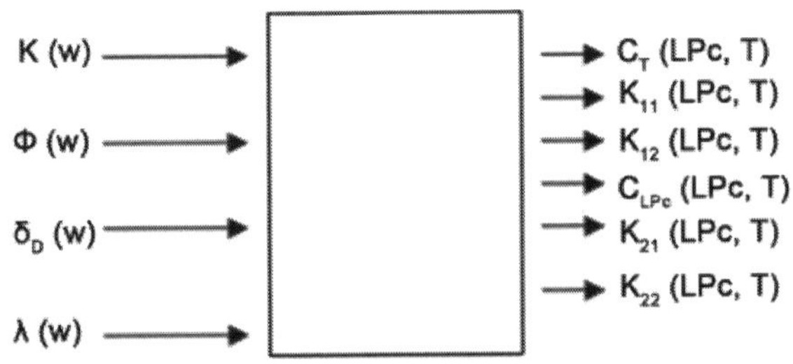

Figure 7. The conversion from measurable material properties into PDE coefficients.

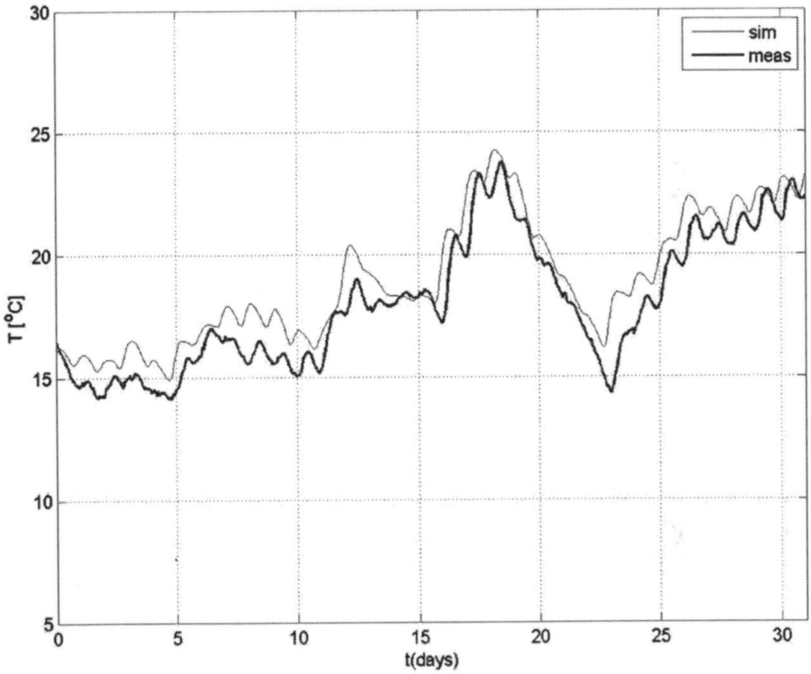

Figure 8. The measured and simulated inside surface temperature.

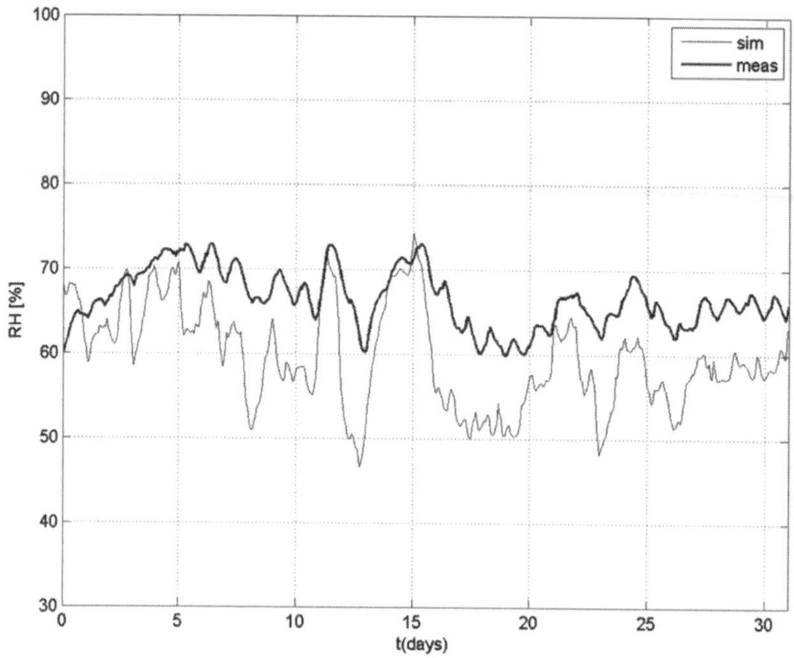

Figure 9. The measured and simulated relative humidity at the surface.

Figure 10. Observed moisture spots at the inner surface near the windows.

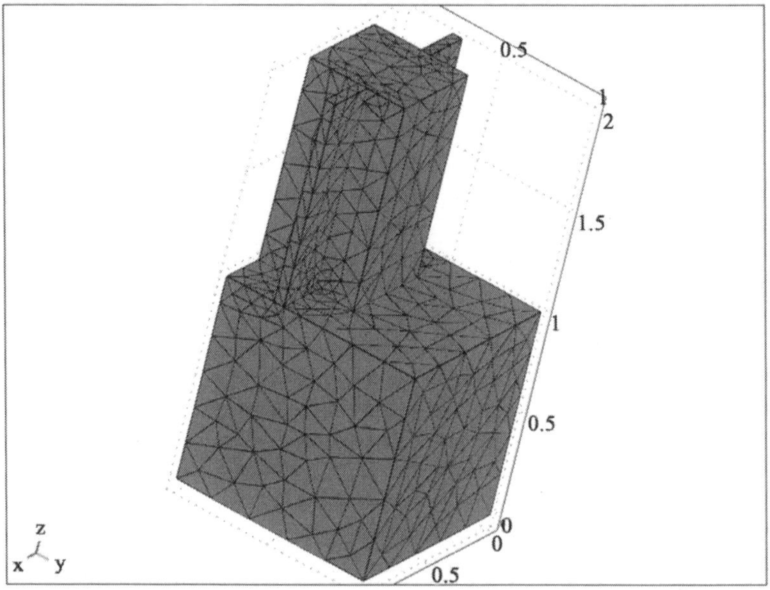

Figure 11. The mesh.

There is no match between the simulated profile at the inner surface of Figure 13 with the observed profile of Figure 10. Therefore it is concluded that the location of Figure 12 is not a possible candidate that causes the observed moisture spots.

In Appendix an overview is provided of nine more moisture profiles caused by corresponding possible wet locations. From these results, the best candidate for the moisture leakage location seems to be at the bottom of the window.

5. DISCUSSION AND CONCLUSIONS

This paper investigates the (in)possibilities of pinpointing moisture leakages from inside surface moisture patterns using inverse modeling techniques.

5.1. Limitations

1) In the presented method, the material properties were obtained from a database (DELPHIN [7]). In order to get very accurate simulation results, the material properties should be measured. This is however quite a problem because, often it is not allowed to take samples from monumental

constructions. So the accuracy of the simulation results is limited. More research is needed to investigate exactly what the effect is of this uncertainty in the material properties.

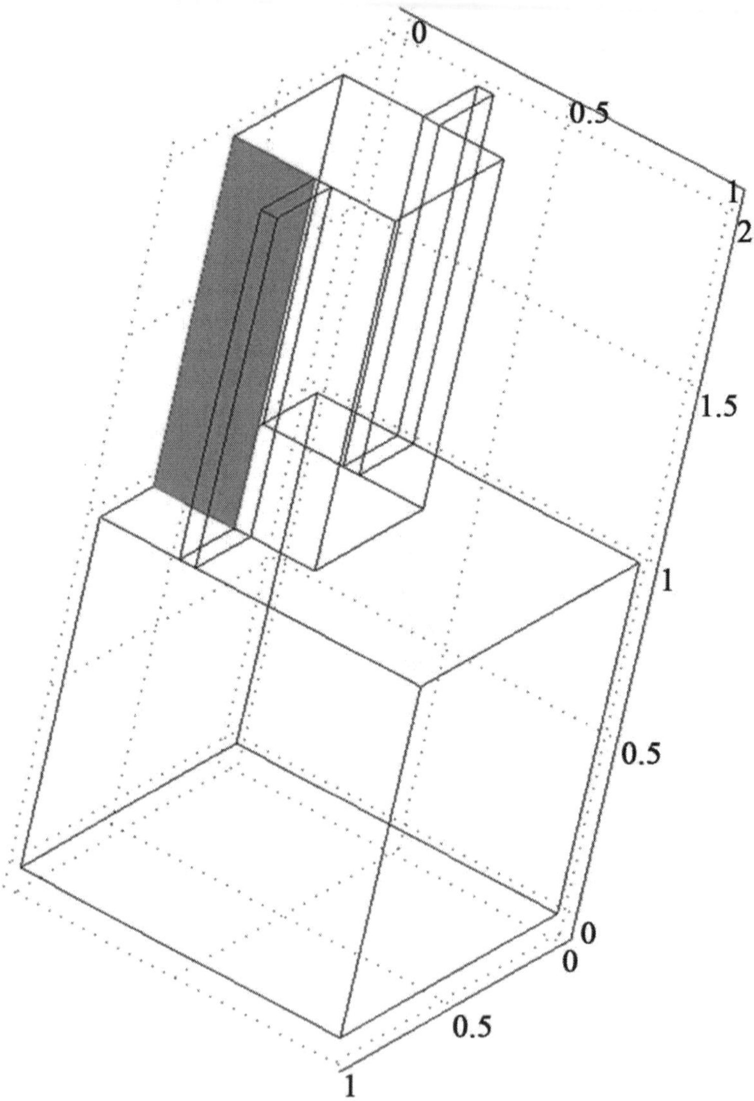

Figure 12. The location of the wet surface condition.

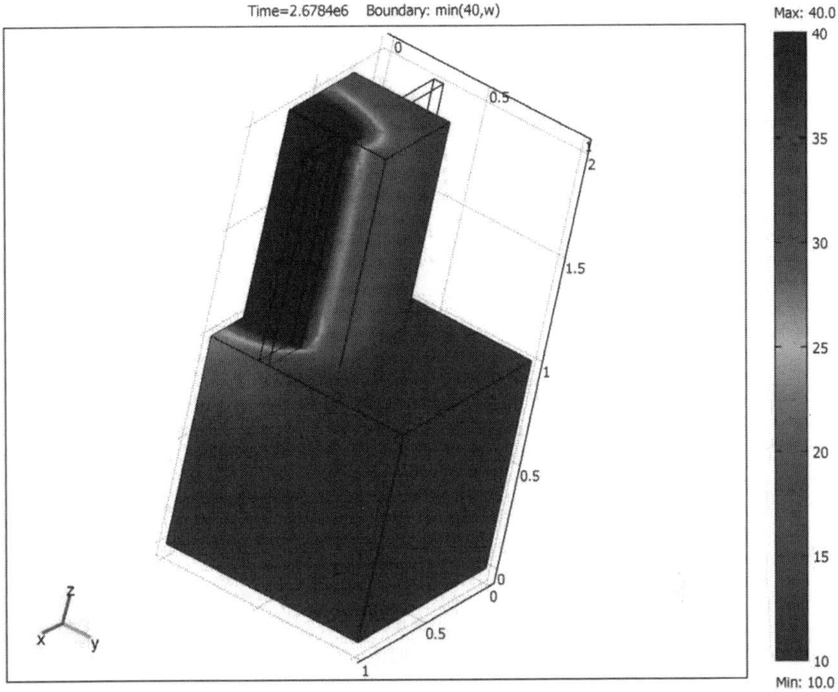

Figure 13. The steady state moisture pattern at the surfaces.

2) The inverse modeling technique tries to match simulated moisture surface profiles with the observed moisture spots. It assumes a single "natural" cause, for example a leak between the window and brick caused by cracks. If a moisture spot is caused by some repair action, for example a complete replacement of the windows, this is not taken into account.

3) The current inverse modeling technique is still rather basic by manipulating the boundary conditions by hand. A more sophisticated method, where the boundary conditions are manipulated by a computer algorithm is under investigation.

5.2. Future Research

Future research include the testing of the method with laboratory experiments and a thoroughly evaluation of the approach as instrument for pinpointing the location of leakages. By selecting materials with well-known properties limitation (1) can be fixed. Also limitation (2) can be fixed because of the high level of control at the lab.

A. W. M. van Schijndel It is concluded that although the presented methodology is promising, more research is needed to confirm its usability.

REFERENCES

1. Briggen, P.M. (2007) A Study on the Moisture Problems at Hunting Lodge St. Hubertus. M.Sc. Thesis, Technische Universiteit Eindhoven, Eindhoven, 120 p. (In Dutch)
2. Briggen, P.M., Blocken, B.J.E., van Schijndel, A.W.M. and Schellen, H.L. (2009) Wind-Driven Rain and Related Moisture Problems at the Tower of a Monumental Building. 4th International Building Physics Conference, 15-18 June 2009, Istanbul, 709-716.
3. Briggen, P.M., Blocken, B.J.E. and Schellen, H.L. (2009) Wind-Driven Rain on the Facade of a Monumental Tower: Numerical Simulation, Full-Scale Validation and Sensitivity Analysis. Building and Environment, 44, 1675-1690.
4. van Schijndel, A.W.M. (2007) Integrated Heat Air and Moisture Modeling and Simulation. Ph.D. Thesis, Technische Universiteit Eindhoven, Eindhoven, 200 p.
5. van Schijndel, A.W.M. (2009) Integrated Modeling of Dynamic Heat, Air and Moisture Processes in Buildings and Systems Using SimuLink and COMSOL. Building Simulation: An International Journal, 2, 143-155.
6. van Schijndel, A.W.M. (2009) The Exploration of an Inverse Problem Technique to Obtain Material Properties of a Building Construction. 4th International Building Physics Conference, 15-18 June 2009, Istanbul, 91-98.
7. DELPHIN (2014) http://de.wikipedia.org/wiki/Delphin_(Software)
8. COMSOL (2014) http://en.wikipedia.org/wiki/COMSOL

INDEX